Following the Legends

A GPS Guide to Utah's Lost Mines and Hidden Treasures

By Dale R. Bascom

Following the Legends

A GPS Guide to Utah's Lost Mines and Hidden Treasures

By Dale R. Bascom

CFI
Springville, Utah

ISBN 13: 978-1-55517-043-5

Published by Council Press, an imprint of Cedar Fort, Inc., 2373 W. 700 S., Springville, UT, 84663
Distributed by Cedar Fort, Inc. www.cedarfort.com

LIBRARY OF CONGRESS CATALOGING-IN-PUBLICATION DATA
Bascom, Dale R. (Dale Rex), 1953-
A GPS guide to Utah's mines / Dale R. Bascom.
p. cm.
ISBN 978-1-59955-043-5 (acid-free paper)
1. Mines and mineral resources--Utah--History. 2. Treasure troves--Utah--History. 3. Global Positioning System.
I. Title.

TN24.U8B37 2007
917.9204--dc22

2007008247

Cover design by Nicole Williams
Cover design © 2007 by Lyle Mortimer
Edited and typeset by Lyndsee Simpson Cordes

10 9 8 7 6 5 4 3 2 1

Table of Contents

Utah

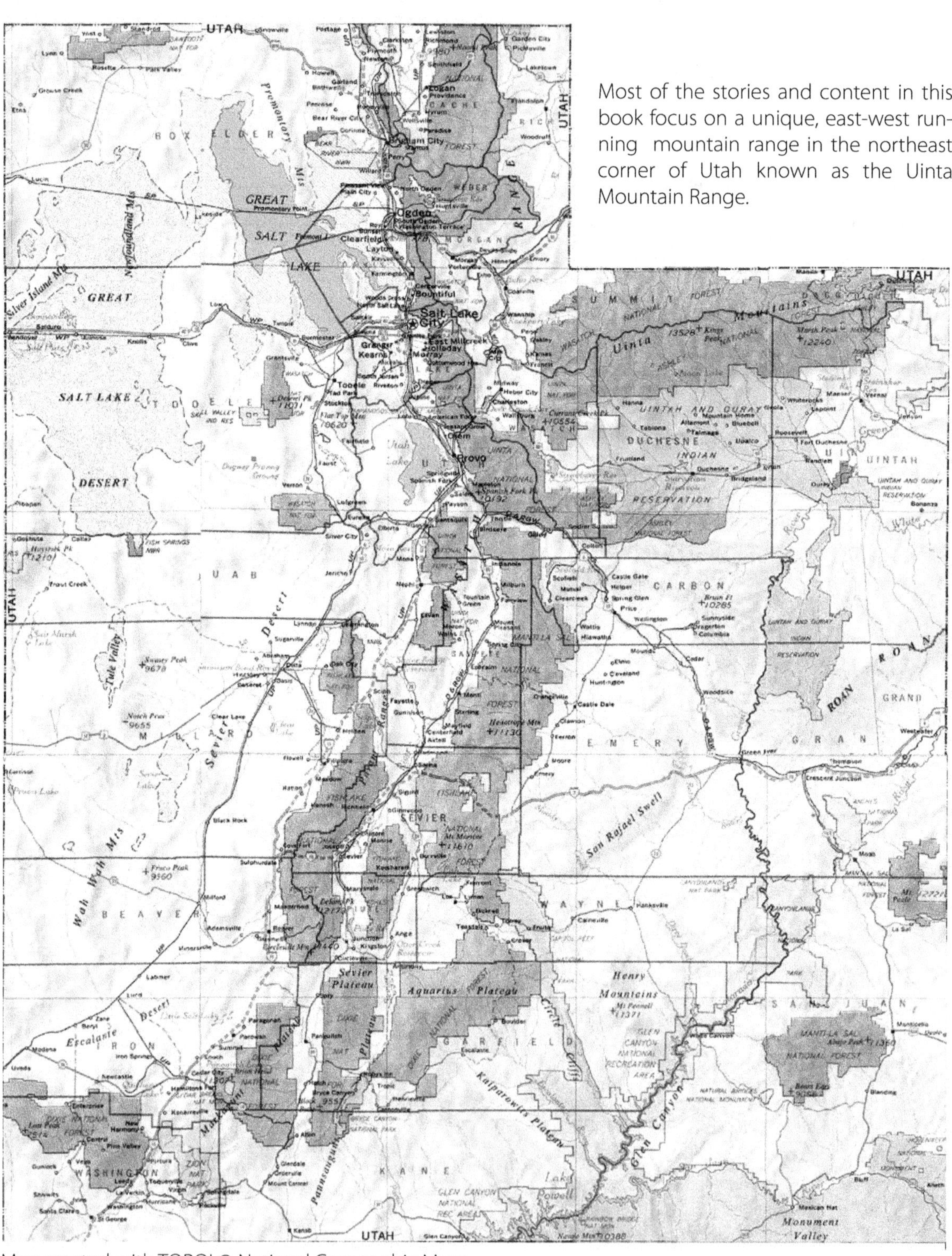

Most of the stories and content in this book focus on a unique, east-west running mountain range in the northeast corner of Utah known as the Uinta Mountain Range.

Map created with TOPO! © National Geographic Maps.

Acknowledgments

Thanks to my wife, Jeri, and our children, who accompanied me to the mountains, wading through ankle-deep water in the early spring or hiking to bug-infested canyon bottoms in pursuit of my dreams. I also thank them for their support, encouragement, and help with this work. Jeri has played the devil's advocate. With her background and expertise in family research, she is all too aware of the importance of documentation. I appreciate her for her support, suggestions, and being my best critic.

Thanks to my friend Cristian Vera for his key contribution to this work. His countless hours in interviews and pencil work to reproduce the symbols are priceless. (See the chapter "More Markers and Symbols.")

In this work I have reprinted pertinent and germane articles from the *Deseret News*, the *Salt Lake Tribune*, and the *Vernal Express*. Permission was granted in April 2006 for the use of these articles. I wish to express thanks to their staff for the reprinting courtesy extended to me. I acknowledge the fact that the information came from them, not from me.

Upon writing to The Church of Jesus Christ of Latter-day Saints to obtain permission to include pictures of a map I call the copper map, I received the reply that they could neither approve nor deny my request because they have no record of the map's existence. The name of the Church is on this document, and it probably originated with the Church.

Occasionally throughout this work I will refer to men known to me only as Aaron and the bishop. These are nicknames Cris gave to two individuals who he met in the mid-1990s in the Provo/Orem, Utah, area. He conversed with them on many occasions and recorded a large number of these conversations on audio tapes. The desire was very strong within me to make the acquaintance of these two individuals. However, for reasons unknown to me, those feelings were not reciprocated. They have both since passed from this life. While I can't meet them, I hope to someday feel closure by visiting their families and feeling through them a bit of the flavor and zest these men had for the great outdoors and for their pursuit of the legends.

Verbal permission was granted by Mick Norton on October 29, 2006, to include any or all of his information in this book. I have learned much from him over the years and appreciate this courtesy.

Paula Rhoades Hepting, daughter of the late Gale Rhoades, granted me permission on October 31, 2006, to include several documents and maps in this work, which originated with her father. I appreciate her permission as well as the excellent documentation her father performed. Some of these maps and documents have been touched up or rewritten to enable you to read them with more clarity. The changes were made while retaining the documents' integrity. I also thank Tammy Hinckley for editing this entire work. Her suggestions are appreciated.

Introduction

It has been said of passion that "there are many things in life that will catch your eye, but only a few will catch your heart . . . pursue these." This describes quite well my outlook on this topic. With the lure of adventure in the wind and two capable feet under me, I feel at home searching for the clues to elusive, yet real, riches that are beyond belief.

Almost everyone enjoys trying to unravel a mystery and I am no exception. In my youth I was intrigued by caves and mines. Often I explored them just to see what was inside. Later in life, while chatting with a colleague named Alton Barzee in 1979, I was informed that his cousin Gale Rhoades had written a book about mines, specifically the Rhoades mines. Almost immediately I purchased a copy. I read it and was hooked on this mysterious legend from then on. Because of my relatively early start in pursuing this legend and associated others, the abundance of evidence and the "people in the know" have been more readily available to me than to some people. This has definitely been to my advantage from a resource point of view. I hope that by sharing the information in this book with you I will pass that advantage on to you.

It's amazing how much has been written on this subject, and it's interesting to contemplate how much more has been passed on verbally, or not passed on at all. Much of the information in this book has come from others, and writing it down is my attempt to preserve it.

I believe I have the privilege and an obligation to recount this historical information and to do it with as much accuracy as possible. With references cited, you may check my accuracy and pursue further research if you have the inclination to do so. As you do, I hope you too will have a desire to convey your findings accurately and with excitement. By so doing, our knowledge of history will retain its integrity. This will solidify a foundation for others to build upon.

Time marches on, and as it does, it is readily apparent that the clues to our past are dwindling. They are fading either through natural forces or from man's neglect or choice. With this concept in mind, I ask that when you find clues to our past, please do not disturb or destroy them, so that others may view and learn of our heritage from them. Thank you. Our heritage belongs to all of us.

This book is written in story form. While camping or at home, my father often lulled my siblings and I to sleep by telling us stories. We enjoyed hearing those stories and we cherish those memories. In this work, as well as on my website (slimsgold.com), I carry on that tradition and offer many stories for your enjoyment.

As you contemplate this work, here is some food for thought: "Truth may often be more colorful and interesting than any fictionalized work."

Implementing GPS Information

All Global Positioning System (GPS) coordinates in this book are given in the NAD 27 Alaska format. As GPS units became available, I recorded my information in degrees, minutes, and seconds. Later I changed to the "degrees, minutes.minutes" format used by Garmin and TOPO. This format is arguably more accurate, but that belief is the basis for the switch. While several manufacturers of GPS units and computer mapping programs accommodate this format, the unregulated version of Google Earth does not. There are only a handful of free computer mapping programs available to the public. Therefore, I apologize for any inconvenience this may have caused. The program I have used to record my waypoints is National Geographic's TOPO program.

The challenge has always been to locate something I've been told about or to return to an exact location where I've already been. With the use of GPS tracking and global positioning, some aspects of the challenge are now out the window. Once something important has been located and recorded in a GPS unit, it may be retrieved and implemented at random. As long as the electronics are reliable and access to the location is available, one can return to the same location. I have been able to get within twenty feet of the location when I have returned.

The other challenge is to find something I've been told about. A hand-drawn map may improve the chances of locating it only slightly. In this work, I have occasionally attempted to give you GPS coordinates corresponding to hand-drawn and dowsed maps. In so doing, I used the TOPO program and gave the challenges my best guess. Therefore I hope you will be understanding if my best guess is off by a considerable distance. My intention was not to mislead you.

At the locations of interest, where I have been and verified my findings, I have offered you the exact GPS readings shown on my Garmin Rino 120 GPS unit. Thanks for your understanding. I hope this information, including GPS coordinates, will be a huge lead for you. Good luck in your search for treasures of yesterday and today.

A Note on Trespassing and Safety

Most of this book deals with locations under the jurisdiction of the Forest Service or the Bureau of Land Management. However, a few items or references will be made to treasures or things of interest on Native American land. Permission to trespass is neither given nor implied in this book. Permission to cross Native American property has occasionally been granted to individuals in the past. The way Native American property is treated now and in the future is reflected in how visitors are treated by the Native American people.

Over the years I have attempted to pursue this hobby (sometimes obsession) with tact, vigor, humor, and respect. Being a family man might have kept me from a few discoveries, but has also kept me from harm's way. I encourage you to also take care as you explore to your heart's content. May your expectations soar and your legs carry you to the mountaintop as you chase your dreams and then return home safe and sound.

Christian Jim's Gold Mine

Author's Note

Marlae Rindlisbacher is a coworker of mine and the daughter-in-law of the late Norris Mike Peterson. She told me this story years ago, and I was very intrigued by it. Then in 1995, she gave me a written copy along with written permission to publish it. I thank her and the Peterson family for sharing it. I also thank another coworker, Liu Toelupe, for editing it. By way of clarification, this story tells of an experience that Hans Mathias had in 1858. Later, his son Ellis Alonzo wrote it down so it could be passed from one generation to another and remembered as Hans told it. Norris Mike Peterson was the grandson of Hans. Stories like this one endure for many generations and keep our dreams alive.

The following story, dictated by Ellis Alonzo Peterson, spins the mysterious tale of a lost gold mine in northern Utah. Many living at the time of these events have attested to the truthfulness of the details of this account. I remember the story well because I heard it told so many times by my father, Hans Mathias Peterson. I shall repeat it to the best of my memory.

I was just a young boy when Christian Jim first came to live with us. We always called him Christian Jim, but lots of people in town referred to him as Indian Jim. He had first come to our house seeking a place to live during the winter. We had a big family, so we didn't have any room in the house. He asked if there was any room in any of our out buildings. We had a room we used for storage in the barn between the harness room and the granary. He looked at it and said he would like to stay there.

There was no talk about payment of any kind. My dad wasn't one to talk about money. He trusted everyone and never much worried about being misused. There had been a lot of trouble with Indians down our way, but Pa always said he trusted this one from the time he first laid eyes on him. We all pitched in and cleaned the place out. Stored there were some broken nests from the hen house, a broken plow blade, some rusted-out containers for brooding chicks, some spikes off an old harrow that Pa was going to sharpen the first chance he got, and a bunch of other less valuable stuff. We moved most of this collection into the harness room. Pa wasn't much at throwing things away. Ma swept the place out and cleaned the one window. Pa and my brother put in a pot-bellied stove that had come from Grandma's house when she died. We got a bed from Aunt Jewel. The mattress was in pretty bad shape, but Ma washed it, sewed it up, and filled it with fresh straw from our threshing pile. Pa made a bench for sitting on and also made a table out of a door that came out of the church when it was rebuilt after the fire.

Christian Jim kept pretty much to himself that first winter. He would fetch wood to the back porch and would see that Ma always had plenty of fresh water from the pump in the backyard. Each morning,

Ma always put a bowl of mush on the water bucket stand by the back door along with a quart jar full of milk. At night she would fix a plate of food just like we had and put it by the water bucket. Always the dishes would come back sparkling clean along with a fresh bucket of water.

In the spring, on a day that Pa was to take a trip into Salt Lake, Christian Jim gave him two small sacks. Each sack was filled with gold. He said one sack was for Ma and Pa for his board, and the other sack was for Pa to turn into cash money for him. Pa was more excited than a hen on hatching day. When Pa got back from Salt Lake, he gave the cash money to Christian Jim and then hurried into the house to tell Ma about the money. I heard Pa announce that the assayer said, "It was the purest gold [I have] ever seen." Pa told us kids to not say anything about the gold.

Shortly after that, Christian Jim left as suddenly as he had come. In some ways, I was kind of glad because I had taken a lot of ribbing from the kids at school because we had an Indian living in our barn. The more I tried to defend him, the more they made fun of me. Finally, I had stopped saying anything. I had already been made fun of because of the clothes Ma had made me along with Pa's silk stove-pipe hat that she cut down for me. After my [friends] finished shooting arrows in it, I never could wear it again.

There were a lot of stories about him around town. Some people said he robbed a bank in the Midwest and was just hiding out here. Others said he killed a man over a bottle of whiskey. Well, I didn't believe any of them. I asked Pa why they called him Christian Jim. I couldn't understand a name like that because he never went to church with us. Pa said it was because of the way he made his mark. Most men who couldn't read or write their name would make their mark, which was generally an X. Then someone who could write would print the name under the mark. Pa said that when Christian Jim made his mark, he would make a cross (+) instead of an X, so folks started calling him Christian Jim. I could tell that Ma missed Christian Jim that summer, and it became my lot to have to fetch the wood and the water.

I tell you we were all smiling when Christian Jim showed up in the fall about the time of the first snowfall. He looked just the same, but this time he had a mule loaded with a pack. He moved right back into his room. That winter we got to be real friends. On a cold winter day, I would slip into his room and talk with him. He told me about the time when he was young like me. He had grown up a long way from here. His family had lived on the plains and hunted the buffalo. When there was buffalo, they were never hungry, but when they couldn't catch the buffalo, everyone was hungry. He said when the buffalos were gone, they would eat their dogs. One day as a boy, he felt sick and couldn't keep up with the family as they followed the buffalos. He was left on a trail close to a stream with some food. A wagon train heading west stopped and picked him up and nursed him back to health. The Wagon Master was heading west to make his fortune in the gold fields of California. As they rode along the way, the Wagon Master told him all about finding gold and how to mine it. When they reached Salt Lake City, Christian Jim left the wagon train and started looking for gold in the area around Salt Lake City. He said for three years he had looked without success, but two years ago he had found what he had been looking for.

Some of the time we talked, he made sacks out of heavy canvas and sewed a drawstring in the top. Other times, he made candles. By spring, he had a lot of candles and a lot of bags made. Just as the year before on a day that Pa was going to town, Christian Jim came to Pa with two small sacks of gold, one for Pa and one for cash money. Pa turned the gold into cash for Christian Jim, and the next day he headed out with his mule loaded with the sacks and candles.

The money from the gold was a great help to our family. The first year we bought a Guernsey milk cow with a heifer calf. The calf looked more like a Jersey. It was small and dark brown in color. Ma said the Guernsey must have been bred by a Jersey bull. It turned out they were both heavy producers and well-mannered animals. We had them for years. The second year, Pa bought a team of horses with a wagon. He got the two blacks with full harness and the wagon. That summer Pa took them to do some hauling for the sugar factory and we made some good money.

That fall, Christian Jim returned again as he did for the next four years. Those were fun winters for me. Always in the spring, there would be two small sacks of gold, and then Christian Jim would disappear for the summer. Once I asked him where he went. He told me that someday he would show me. One spring I saw him talking with Pa. Pa nodded his head. That night Christian Jim told me he was going to show me where he went during the summer if I wanted to go. I was so excited I couldn't sleep that night. The next morning Ma had some food ready for us. Pa saddled up the buckskin mare with the army saddle that he bought when Johnson's army pulled out. I rode the buckskin and Christian Jim walked, leading the mule.

When we got north of town, just past Peter Massenpeel's house and the blacksmith shop, Christian Jim stopped and said that where we were going was a secret and he wanted to blindfold me if that was all right with me. I said, "Sure." He had one of his canvas bags that he had cut a breathing hole for my mouth. I put it over my head. It was sure dark in there. We walked around in circles and then headed out. At noon we stopped in a grove of trees to eat and rest the animals. I took the bag off while we ate. After about an hour, we started again with the bag back in place on my head. In the afternoon the sun beat down on my head and it really got hot in that bag, but I didn't complain. I could feel the sun on my right side and back so I think we were traveling north. That night we stopped near a creek, camped and picketed the buckskin and the mule, ate dinner, and rolled our blankets out.

The next morning we were on our way early, with the bag back on my head. By now we were following a stream of water and we were moving uphill. We stopped often to let the horses rest. About midday, we turned right up a steeper trail. I could tell the stream was smaller now. Soon we stopped to rest the animals and had lunch. We were in a small narrow valley with a stream about three to five feet across. After lunch Christian Jim took me over to a rock and told me to look on top of it. There on the top was a cross that had been chiseled into the surface of the stone. The stone was about three feet long and about two feet wide and it was on the right side and close to the stream. The cross was about twelve inches long and about six inches wide. He said he had made a series of these crosses that would be markers to lead him to his mine.

We continued up the canyon. I was no longer required to wear the bag on my head. From time to time, Christian Jim would show me another cross that was chiseled in a rock. Each time the cross was turned a different way. He explained that with each cross you had to take your direction from a different point of the cross. The first cross, you follow the point at the top of the cross. The second, you take your direction from the right arm of the cross. The third time you get your direction from the base of the cross. The next time, you get it from the left arm of the cross and so on around the cross with each succeeding sign. Most of the crosses were near the stream until we got to the top of the little canyon. By mid-afternoon, he showed me the last cross. It was up on a side hill. He asked me to sight along the base of the cross to the tip, and asked me what I saw. I said, "Just a clump of bushes." We went back down to the stream and found a shelter hidden in some trees. The shelter had been made of willow branches with fir boughs placed on the top to keep the rain out. That afternoon we rebuilt the shelter and put new boughs on it. Christian Jim disappeared and returned with a pot, an iron frying pan, a grill for a fireplace, and some other things to make a comfortable camp. We ate dinner and turned in. It had been a long trip and we both were tired.

The next morning Christian Jim asked me if I could remember the clump of bushes that the last cross pointed to. I remembered and walked up the hill right to the spot. He asked if I could see anything strange here. I said, "No."

"Follow me," he said. He walked around one large bush and pulled back the branches. There was a hole in the side of the hill. He motioned for me to follow him, but said to be careful because sometimes there are rattlesnakes in there. He disappeared into the hole and I followed. It was dark, and I felt my way along the walls. A light appeared in the tunnel and I saw Christian Jim lighting candles. The tunnel widened into a room. It was still dark but soon my eyes adjusted and I could see a room about thirty feet long and ten feet wide. Along one side was a pile of loose rock and stone. On a shelf was a pile of chisels, hammers, an iron bar, a pick, and a shovel that leaned against the wall. Back in the corner stacked on another shelf was a pile of canvas bags like those Christian Jim had stitched in his room. These were bulging from the bottom to the draw string at the top.

As the candles flickered, there were sparkles all around the room. Christian Jim told me the sparkles were quartzite, a hard rock that imprisoned the gold. He showed me veins of gold that were captured in the quartzite. At the end of the tunnel about the height of my overalls breast pocket, he showed me a vein of gold about as large as the handle on Ma's butter churn. He told me that he had been following that vein for about forty feet. Sometimes it may get as large as a woman's wrist, and sometimes it gets as small as his little finger. He had to chisel the stone away to get at the vein. Sometimes he would follow a fork in the gold vein until it ran out, then he would return to the main vein. Any chips that had signs of gold in them, he placed on the pile on one side of the cave. When the vein ran out, he planned to crush that rock to retrieve the gold. The rock that had no gold, he carried down and dumped it in the stream to be washed away. He wanted no tailings pile to give away his secret.

Christian Jim opened one of the bags and let me look in. It was filled with pure gold. I opened a second sack, pure gold, a third sack, the same, pure gold. I lifted the sacks and they were heavy for their size. I judged each weighed about as much as a water bucket filled with water. I couldn't see for sure, but there must have been 20 or 30 bags. Christian Jim gave me a chisel and a hammer and told me I could have all the gold I could take out that day. I started working feverishly. It wasn't long before my wrist was so tired that I could hardly lift the hammer. My knuckles had been busted from the hammer missing the chisel. At the end of the day, my body was hurting all over. My sack had only a few broken chips of the precious metal. It was only then that I realized the magnitude of Christian Jim's work in cutting away this cave to give up its prize.

Christian Jim had set some snares the night before. When we got back to camp he checked the snares and came back with a rabbit that was soon on a spit over the fire. I can't remember eating anything that tasted so good. As we sat around the fire that night, Christian Jim told me that that was his last summer. This fall, when he came down, he would come and get me and together we would carry the gold out. He was then going back to the plains to find his family. He said he wanted to do some good things for Ma and Pa before he left and that he wanted me to have the mine. I couldn't sleep that night;

I was so excited thinking about all that gold. I wondered about how far that vein ran and how much gold was left there.

The next day, early, I climbed up to a big circular outgrowth of rocks on the side of the hill and looked down at the clump of bushes that hid my gold mine. After breakfast we started back with me on the buckskin mare and Christian Jim leading the mule. When we got to the stone with the first cross, I put the bag on my head again and we retraced our steps back down the trail. We didn't stop much now. When it grew dark, I fell asleep in the saddle. Towards day break, we stopped and Christian Jim took the bag from my head. We were back at the church in Lehi.

After breakfast, Christian Jim said he was starting back. I walked to the gate with him. He told me to help my ma and pa to be ready to help him fetch the gold out this fall. He clasped my hand, looked into my eyes, and turned and headed up the road leading his mule. That was the last time I ever saw Christian Jim. No one else ever remembered seeing him after that day either.

As fall approached, I kept watching up the road hoping to see my friend Christian Jim leading his mule back to our place. Pa asked around that winter and the next summer, but no one ever remembered seeing Christian Jim. Pa asked the merchants and bankers in Salt Lake if they had seen any signs of gold coming into town. The only signs were a few nuggets that were coming back from California. The assayer in town said that the only pure gold he had seen was that which I turned in from my hard day's labor. Even though it was a small amount, it went a long way to helping me on my mission.

When I returned from my mission, I took every chance I could to retrace my trip to Christian Jim's mine. I thought I could remember that trip but the mind plays funny tricks on you. I explored every canyon within two days of our home. I tried to recall the sounds and the direction of the trip, but no matter how hard I tried, I couldn't find the right canyon. Every time I thought I was on the right trail, I would end up on unfamiliar ground. My eyes were always searching for the rocks with the Christian Jim chiseled crosses. I know if I could find a cross, I could find the mine, and perhaps I could find the remains of my friend and put him in a proper resting place. I don't know what happened to Christian Jim, but one thing I do know. He wouldn't have left without saying good-bye to me, and Ma and Pa. I often wake up at night and think of my trip with Christian Jim. If only I could find one of those chiseled crosses.

Sugar Loaf's Golden Treasure

Author's Note

Vern Bullock gave me a hard copy of the following story in the 1980s. In September of 2002 permission was obtained from his sons Neil and Dave to place it on my website and in a book. I thank them for this courtesy. I also thank Dave for his editing suggestions.

A tunnel filled with gold ingots, some covering the body of a dead trapper, drew like a magnet on Johnny Rasmussen's mind for fifty years to lead him from Old Mexico to the state of Utah in the USA in the year 1913. (The year may be incorrect.) Thinking back when paper and pen were available, Johnny's great-grandfather listed down the things most essential to guide him back to a treasure of gold ingots piled over the body of his dead trapping partner and many dozens more still piled neatly along the walls of the ancient tunnel. The description of the treasure location was:

1. Two and one half day's ride from the south end of the salt sea and follow a river which [runs] northward from a large fresh water lake located in a beautiful valley southerly from the salt sea.
2. Half a day's ride southerly from the eastern shore of a fresh water lake to a sugar loaf peak.
3. The sugar loaf peak is south easterly above an area with many springs that make a valley at the foot of the mountains and supply an Indian camp with water.
4. The gold tunnel is about three quarters of a mile south of the sugar loaf peak and high on the foothills.
5. The sugar loaf gold tunnel is below some rusty red ledges.
6. You look to the southwest along the mountain and the valley closes.

With these guide posts and a rough way of directions Johnny's great-grandfather was sure that some day he could return to the hidden sugar loaf gold treasure. But marriage, making a living and settling in Old Mexico consumed the years and old age and death ended the life long desire of Johnny's great-grandfather to again find his treasure. This was all brought about in about 1825 when Johnny's great-grandfather was with a trapping party in the Rocky Mountain area.

It was getting along to early fall. Fur bearing animals were becoming scarcer so the trapping party held a powwow in a meadow southerly from the south end of the salt sea. (The family thinks the Grantsville area.) This search was for future trapping streams. It was decided to separate the party into groups of three and spend about ten days to two weeks fanning out in all directions to thoroughly explore the areas for the next spring's trapping.

Rasmussen and two companions were sent in a southerly direction following a large stream. For two and one half days they explored branching streams but kept to the course of the starting point stream as a guide line. When they entered a mountain nearly southerly from their starting point they beheld a beautiful valley with a large fresh water lake south of them. In this valley a number of streams flowed from the mountains to the east which they explored but found no evidence of beaver dams. To the south the valley closed so they engaged in following streams as far south as the canyons led. One little stream had a beaver dam in it and they followed its course southerly toward the mountain. A sugar loaf stood out at the base of the mountain for which they were heading. They made their way through cedar trees in the valley and saw the Wickiup of an Indian camp westerly from sugar loaf. Not knowing weather the Indians were friendly or not they skirted to the north of the camp and headed for the hills. The stream they had been following was fed by many springs between the Indian camp and the mountain.

When southerly from the camp and ascending the foothills they heard and saw a party of Indians coming from the south. They forced their horses higher up the hills and through the cedar trees. One of the Indians sent an arrow which lodged in the back of one of the trappers. He hung doggedly to his saddle until they saw a badger come out of a large hole on the side of the mountain. One rider jumped from his horse and found the hole to be large enough for them to crawl down into and then a larger opening beyond. They abandoned their horses by swatting them on their rumps. The trappers hurried down into the badger hole, pulling the wounded man down into the hole with them.

It was now evening time and getting very cool. A pile of sticks and brush had been dragged into the hole by animals and from these eventually a fire was started to warm the wounded man. One knotted cedar branch made a good torch and was taken from the fire to light the cavern to its depths. Not far along the way one of the trappers found piles of bricks which proved to be gold ingots. He hurried back to his friends with the news of his find and the two men ran back to look at the gold. They were very excited with the great treasure. After some discussion it was decided that the gold bars were too heavy for one man to carry without horses and the gold could not be continually hidden. When found out they had it their lives would be in danger until they told where their wealth came from. They came to the decision that the best thing would be to keep in mind the guide posts that would lead to their find until paper and pen could be had to draw a map and write a thorough description which would lead them back to the sugar loaf gold.

This decided they hurried back to their wounded friend to tell of their good luck but found he had died from his wound. They pulled his body back to the piles of gold bricks and completely covered his body with bricks of gold to keep wild animals from ravaging him. From fear of being discovered by the Indians the trappers waited for days before they dared venture outdoors.

The hour was late when they crawled out of their treasure tunnel and walked the long distance back to their home camp on the shores of the salt sea. It is not known how long Johnny's great-grandfather stayed with the trapping party but he kept his golden tunnel treasure of the sugar loaf a secret and firmly in his mind until he had paper and pen to make a permanent record of it.

[At this point, the narrative changes voice for quick interjection.] The writer (Benjamin Vern Bullock) does not know much about Johnny Rasmussen's background or anything about his family. He does know of his coming from Old Mexico in his late fifties and settling in Springville, Utah to search for, and hopefully find, his great-grandfather's treasure. His great-grandfather eventually wrote the description of the treasure and his story about discovering it on an animal hide, which is the same hide that Johnny had with him when he came north to look for the treasure.

Johnny guarded the map selfishly and pondered maps of the United States to learn of its location. The only salt sea in the Rocky Mountains was the Great Salt Lake. The stream coming from the south which they followed was known as the Jordan River. It comes from the fresh water body known as Utah Lake. The city of Provo was easterly of the center of Utah Lake and the town of Springville was six miles southerly of Provo and a good place for his headquarters from which to explore.

Johnny told some new acquaintances why he had come to Springville and showed them the map. One old boy became excited and said, "Johnny you have come to the right place. There is an old fellow south of here who has a dream mine. I think it's what you are looking for. He was shown nine rooms filled with gold bricks and ancient records in a dream and according to your record it is the exact location." A few days later Johnny's story had reached the miner John M. Koyle, of Leland, Utah. Leland is a little farming community just south of Spanish Fork, Utah. Johnny was invited to the Koyle mine and there, in the cabin of the new dream mine workings, told his story. After he finished, John Koyle, the owner of the mine, summed it all up and said, "Johnny, you are in the right spot. The way your great-grandfather traveled, exploring streams and heading south, it would have taken him about one and a half days to climb up the foothills from the easterly shore to where we are now. This canyon is the north side of the pyramid or sugar loaf mountain. Below you where Salem pond is now used to be springs. If you look to the south the valley closes so Johnny you are home." Johnny was a little confused and said his great-grandfather rode through cedar trees in the valley and on the foothill. Koyle explained to him that the pioneers or early settlers in the valley cut down the cedar trees for fuel and fence posts, denuding the area. Johnny was all but converted, and then he remembered and asked Koyle if he had any rusty ledges. Koyle said they didn't but the tunnel was started in a ledge and they had an iron blowout high on the mountain. This didn't satisfy Johnny and he had to ponder over the location of the dream mine to consider whether it was the location for which he sought. It just came too easy to be brought to where someone else was seeking the same treasure.

Lars Olson, the superintendent of the dream mine, sat in on the meeting. When Johnny asked about the rusty red ledges it rang a bell in Olson's memory. He had a neighbor in Provo, Utah where he lived who owned a little mine fifteen miles south from Koyle's mine. Ben Bullock, the owner, had some red ledges. The whole location of the two mines fit to a tee; the story in Johnny's description with the exception of the rusty ledges which Mr. Bullock's location had and Koyle's didn't. Also on a rise above the springs by Bullock's location were the camping grounds of old Chief Blackhawk. It may have been one of his warriors that put an arrow into the back of the trapper. When Olson returned home for the week, he contacted Bullock and told him Johnny's story. This led to locating Rasmussen and arranging a trip to Bullock's "Golden Relief Mine" just south of Spring Lake, Utah. This little lake had been made by building a dam to back up the waters of the springs which were below the Indian camp. On a Monday Bullock hitched his team to his three seated buggy and with Olson collected Johnny for the trip to the Bullock mine.

Bullock had two mines. One drift was in a little canyon below some rusty red ledges. The other one was a long tunnel high on the mountain in yellow rock canyon known as the Syndicate Mine. As the party rode south out of Payson Mr. Bullock pointed to the mountain farther south ahead of them and said, "This is your sugar loaf peak. The springs are a little north and below it is where a little lake is now. Chief Blackhawk had his camp on the rise above the springs. About three quarters of a mile south is my little mine. Above it a couple of hundred feet are some rusty ledges. If you look to the south you will see that the valley closes and if you look to the right you will see Utah Lake. Your map fits the dream mine exactly except for the red ledges.

When the party arrived at the ranch below Bullock's "Golden Relief Mine," Johnny acted a little disgruntled. He looked at the valley and foothills and saw no cedar trees. Mr. Bullock told him the same story as Mr. Koyle had about fencing and burning the trees for fuel. Johnny saw hundreds of cedar fence posts dividing every parcel of land that was under construction. Mr. Bullock unhitched the team and threw a saddle on one horse and a pack saddle on the other. He filled the pack saddle bags with picks, shovels and grub. He then swung into the saddle and led the pack horse. Those that wanted a lift grabbed the horses' tail as they started toward the mine. They found the miners eating lunch so Mr. Bullock made a hurried trip to the working face after which they went up the canyon to the rusty ledges. The horses were tied to oak brush. Something which looked like an old mine dump was found behind some brush and below the red ledges.

All of the men got excited and with enthusiasm began excavating the dump level. They brought in teams, plows, scrapers and other equipment. They spent a week trying to find the lost tunnel entrance. Johnny was so excited he expressed his desire to obtain a tent, clear a spot on the old mine dump, and move his residence to the site with Bullock bringing him supplies and water.

The next two days were spent in getting equipment collected for the treasure hunt. Thursday afternoon four loaded wagons arrived at the old Flanders ranch and a trek was started to deliver excavating equipment to the lost tunnel site. Scrub oak trees were pulled out by the roots, sage brush grubbed out and work went ahead leveling an area about one hundred feet long on the level with a plan to go toward the mountain to bedrock.

The second day of excavating brought the most excitement of the project. A plow point hit a cedar tree stump about eight inches in diameter. The roots were still firmly imbedded in the soil and were well preserved. The stump stood about five feet tall and was left as a happy reminder to Johnny that his great-grandfather had truly ridden through cedar trees. About eight feet further into the hill a pile of charcoal was found. This the men sacked up to use for black smith purposes. This charcoal find led to the story of finding an old slag dump several years before. It was located between the dream mine and the Bullock properties. It was evident that smelting of ore had been carried on in the vicinity during ancient times. It is believed that this work may have been done even before the time of the Spaniards. It has also been learned since then that back in those "ancient" times, mining was done by building a fire in the face of the tunnel heating up the rock. Then, cold water was poured on the rock causing it to fracture which made it easier to remove. Perhaps this is the reason for the stockpile of charcoal.

As the work progressed of looking for Johnny's treasure tunnel Mr. Bullock's miners were making footage in their tunnel and a fissure producing a nice supply of water was struck. With water being available just a few yards from the excavation project there was no need to take horses off the mountain each night. Baled hay was brought up and the camp centered below the rusty red ledges on the newly excavated site.

After two weeks of work no evidence of Johnny's treasure was seen. Mr. Bullock decided upon a different approach which met with the favor of all. He moved his miners to the new site. They then drifted into the mountain fifty feet or more then they would make a right turn and skirt the contour of the mountain at depth for the ancient workings. This would also prospect the formations to the south. The work was started and Johnny stayed on until the cold weather set in. His belief in the site was firmly founded by the discovery of the cedar tree stump still solidly rooted in the mountain soil.

During the winter Johnny left Springville. He wrote Mr. Bullock several times giving him encouragement and hope of finding the old tunnel filled with the ingots of gold. The letters stopped coming and Johnny was never heard of again. His great-grandfather's map and history of its location were gone with him. The excitement died down about finding the sugar loaf gold hoard.

Mr. Bullock continued prospecting in the new golden relief tunnel and worked intermittently until 1923 when work was stopped at a distance of 900 feet. At this point there was a nice stream of water containing an iron oxide. In 1955 Mr. Bullock's son Vern revived work on the old site one hundred feet to the north. This tunnel was used to store water tanks for use at the foot of the mountain. Vern entered the old tunnel and found the oxide in the water had formed stalactites and stalagmites. They were so close together that they had sealed the last fifty feet of the tunnel. The portal had partly filled with soil and the water had backed up with a deposit of gel like oxide mud a foot deep in the bottom of the drift. The acids and heavy mineral content had eaten the rails until just a thin rusted skeleton of steel was left.

Again in 1975 to 1980 more work was done in search of gold-silver ore. Formations are very promising and work is continuing but Johnny's lost gold hoard has never been found. The trapper's remains are still safely protected by the cover of gold ingots which keep him from ravaging animals since about 1825.

B. Vern Bullock: dedicated to my three sons, Neil, Hart and David and my favorite daughter Beverly; all of whom I enjoyed lulling to sleep with such stories as this.

The Mythic Aztlan

Author's Note

The next eight chapters are excerpts from newspapers. I have included them as a fast and easy reference. I hope you enjoy these stories and will use them as a database for further research into Utah's mysterious legends.

The Salt Lake Tribune

Bits of History Suggest Utah Is Location of Mythic Aztlan

Nov. 17, 2002
Tim Sullivan

It was a map drawn in 1768 by a Spaniard in Paris that sent Roberto Rodriguez running toward Aztlan.

As a Mexican American, Rodriguez long had pondered the historical location of Aztlan, the mythic homeland of the Aztecs. Six years ago, he and his wife, Patrisia Gonzales, found tantalizing directions in Don Joseph Antonio Alzate y Ramirez's map of North America.

Where present-day Utah would be, and next to a large body of water called "Laguna de Teguyo," are the words: "From these desert contours, the Mexican Indians were said to have left to found their empire."

That cryptic message is one clue among many—a petroglyph etched on a sandstone wall in eastern Utah's Sego Canyon, an 1847 United States map highlighting the confluences of the Colorado, Green and San Juan rivers in southern Utah, a mound and more petroglyphs just outside Vernal—that have researchers considering a new angle on the history of the southwestern United States.

"Some don't believe [Aztlan] was true, like Atlantis or the Garden of Eden," says Roger Blomquist, a doctoral student at the University of Nebraska at Lincoln. "But I'm convinced it's in Utah. The evidence is very compelling. It's building a mosaic that supports that thesis."

Since the 1960s and '70s civil rights movement, Chicano activists have used the name Aztlan to describe the American Southwest as a northern homeland for Americans of Mexican heritage. But for much longer, people all over the world have been trying to pinpoint the historical location of the legendary place the Aztecs left to build their civilization in the Valley of Mexico.

Rodriguez says Aztlan's literal and figurative meanings are both relevant to his search.

"People would always tell us to 'go back to where we came from,' " Rodriguez says. "Then we came up with this map. Our work is about whether we belong or not."

Western scholars, Catholic clergy, Chicano activists and even the Aztecs themselves have been seeking Aztlan for more than 500 years. They have put much of their energy into gleaning facts from the story that tells of a people emerging from the bowels of the earth through seven caves and settling on an island called Aztlan, translated as "place of the egrets," or "place of whiteness."

Acting upon a command from a spirit, these people left Aztlan and went south until they came upon an eagle devouring a serpent in the present-day location of Mexico City, where historical records suggest they founded the city Tenochtitlan in the 14th century. But in 1433, Aztec leaders burned the picture books that recounted the migration to the Valley of Mexico, leaving only oral tradition and the name Aztlan.

The Aztec king Motecuhzoma I was probably the first to investigate seriously the location of Aztlan. In the 1440s, he sent 60 magicians north for a journey that itself became a legend—according to chronicler Diego Duran, these pilgrims encountered a supernatural being who transformed them into birds, and they flew to Aztlan.

After the Spanish conquered the Aztecs in the early 16th century, they began studying the Aztecs' origins. Francisco Clavijero, a Jesuit priest, in 1789 deduced that Aztlan lay north of the Colorado River. Other Mexican, European and American historians put Aztlan in the Mexican state of Michoacan, Florida, California, even Wisconsin. Many others deny it ever existed.

But perhaps the most widely accepted historical location of Aztlan is that proposed by historian Alfredo Chavero in 1887. Retracing Nu-o de Guzman's 1530 expedition north from the Valley of Mexico, Chavero deduced that Aztlan was an island off the coast of the Mexican state of Nayarit called Mexcaltitlan.

"Some don't believe Aztlan was true . . . but I'm convinced it's in Utah."

Modern-day scholars who favor Utah as an Aztec homeland use some of these studies and chronicles to advance their theories, which range geographically from Salt Lake Valley to the Uinta Mountains to the Colorado Plateau. But each of these researchers also seems to have his or her own trump card.

Rodriguez's curiosity originally was spurred by a copy of an 1847 map of the boundaries drawn by the Treaty of Guadalupe Hildalgo, but quickly expanded to "a hundred others," including the chart Alzate y Ramirez created for the Royal Academy of Sciences in Paris. The maps touched off "Aztlanahuac," a project by Rodriguez and Gonzales, newspaper columnists whose work appears in The Tribune that has spawned one book with two more on the way.

Aztlanahuac led them to gather oral histories on migration from Native Americans throughout the Southwest. Believing that the "Laguna de Teguyo" had to be the Great Salt Lake, the San Antonio couple also traveled to Antelope Island four years ago. There, Rodriguez asked a state park ranger how many caves the island had. The ranger's reply was, of course, seven.

Blomquist, a doctoral candidate in American Frontier History whose dissertation explores Aztec origins in Utah, focuses on the Uinta Mountains. He believes that Aztecs, who would have heard ancestral stories, advised 17th-century Spanish prospectors to look for gold in northeastern Utah.

Blomquist also cites a "natural temple site" in the Uintas near Vernal. He says there is a 200-foot-high mound with footsteps carved into it and an altar-sized boulder at its base that mirrors temples he has seen in Mexico, such as Monte Alban outside of Oaxaca.

On a rock at the site are petroglyphs of a warrior and his family that Blomquist says doesn't resemble rock art of the Fremont people known to have inhabited Utah. And the warrior is carrying a long sword-like object that broadens to a blunt end, like a cleaver, which Blomquist likens to a Mesoamerican weapon called a macana.

Then there is Cecilio Orozco, a retired California State University at Fresno education professor who has observed that petroglyphs in Sego Canyon, about 30 miles east of Green River, correspond to the Aztec calendar's mathematical formula of five orbits of Venus for every eight Earth years. On one of the canyon's sandstone walls are two petroglyphs of knotted string, one with five strings hanging down, the other eight.

In conjunction with his mentor, Alfonso Rivas-Salmon, Orozco theorizes that southern Utah is not Aztlan but the earlier homeland of "Nahuatl," the land of "four waters," where the Colorado, Green and San Juan rivers meet to pour through the Grand Canyon (Nahuatl is also the name of the Aztecs' language.). The 1847 treaty map also points to southern Utah as the "Ancient Homeland of the Aztecs."

Along those lines, Belgian scholar Antoon Leon Vollemaere believes he has pinpointed the location of Aztlan on either Wilson or Grey Mesa, where the Colorado and San Juan meet under Lake Powell.

Researchers also cite the close connection between the languages of the Aztecs and the Ute Indians in the "Uto-Aztecan" linguistic group, as well as the coincidence that the Anasazi culture began to decline at about the same time the Aztecs' ancestors were supposed to have left Aztlan.

While the pile of evidence that the Aztecs came from somewhere in Utah may seem high, more skeptical scholars like Northern Arizona University archaeologist Kelley Hays-Gilpin put things into perspective.

Hays-Gilpin acknowledges the linguistic connection between the Aztecs and Utes as well as economic interaction between Mesoamerican and North American peoples. But she offers a twist on the overall migration scheme—the Aztecs' ancestors may have moved north before moving south.

Hays-Gilpin believes that people speaking a proto-Uto-Aztecan language domesticated maize in central Mexico more than 5,000 years ago and consequently spread north to an area of the American West that could have included Utah. Out of that multitude of cultures, some groups could have migrated south to northern Mexico, and some of those could have, as she says, "moved to the Valley of Mexico and subjugated some of the confused and bedraggled remnants of the latest 'regime change.'"

This concept resonates with Utah Division of Indian Affairs Director Forrest Cuch, a member of the Northern Ute Tribe, who remembers his grandmother telling him his people came from the south. Could the Utes and the Aztecs' ancestors also have lived in close contact in modern-day Utah?

"I'm open to it," Cuch says, "because so little is known about the past."

As such, it would be almost impossible to prove the historical location of Aztlan, but Roberto Rodriguez says clearing the mist surrounding the myth may not be so important anyway.

While treading the path of his Aztlanahuac project, Rodriguez began to uncover a history of mass migration akin to the one Hays-Gilpin suggests. For him and Gonzales, understanding the larger scheme of historical movement throughout North America became more vital than deconstructing one elusive origin story.

"[Finding a location] has almost become irrelevant," he says. "Now, we have a bigger understanding, that the whole continent is connected. You have all these stories of people going back and forth."

Rodriguez says all that migration is most significant for Mexican Americans, and for the thousands of people now moving from Mexico to the United States, because it affords them and subsequent generations an answer when someone says, "go back where you came from."

"I just hope kids at school some day will at least be shown these maps," he says.

University of Utah ethnic studies professor Armando Sol-rzano has tailored the Aztlan concept to fit Utah, which is experiencing its own influx of Mexican immigrants.

Sol-rzano, a native Guadalajaran, has his own reasoning as to why Utah was a point of departure for the Aztecs—that the geographical characteristics of Salt Lake Valley resemble those of Mexico City—but his interpretation of Aztlan is, like Rodriguez's, a broader one.

Sol-rzano tells of arriving in Utah 12 years ago and seeing the Wasatch Mountains and the Great Salt Lake. "I said, 'my [gosh], this is Aztlan.' I felt a spiritual unity with the land, something I had never felt before outside Mexico."

He compares the concept of Aztlan as a sacred land of harmony with that of Zion in the Mormon tradition. The similarities, he says, show that both cultures are searching for a common goal. Sol-rzano calls his Utah adaptation of Aztlan "Utaztlan."

Had Sol-rzano's own migration path taken him to a different part of the United States, his concept of Aztlan likely would be different. Still, he shares his sense of the myth's importance with people of Mexican heritage all over the country.

"What is happening now is we are returning," Sol-rzano says. "This is an opportunity to rewrite history and make justice."

The Uintas: The Mother of Gold

The Salt Lake Tribune

Madre de Oro del Uinta

May 7, 1950

James P. Sharp

Whether you believe this tale or not, you'll find it fascinating. If you believe it, you'll have a difficult time restraining yourself from leaving today for an extended trip into the Uinta basin.

If historians write correctly, then one Francisco Vasquez de Coronado, who was a high Spanish official, left Compostela, Mexico, Jan. 1, 1540. He was at the head of an expedition that went into what is now New Mexico, searching for the Seven Cities of Cibola, where the houses were said to have been built of solid gold.

Failing to find these mythical cities, he dispatched one of his lieutenants, Garcia Lopez de Cardenas, with 12 men, and told them to search for those cities and for the gold to the north and west. A definite date and place were agreed upon for the two parties to reunite.

Just where Cardenas and his men went we do not know, but Young, in his book, *The Founding of Utah,* states they might have reached the eastern part of what is now Utah. Other writers make mention of the same fact. Just where they went, what they found or what they did seems shrouded in mystery. Now let's jump about four centuries.

Late Summer, 1922

In the late summer of 1922 a tired team hitched to a buggy stopped at my place on the upper Provo. What I took to be five Indians were riding in it. One asked permission to camp in the pasture for the night. They were gone when I arose the next morning but about two weeks later they returned. They said about as much as they had on their first visit and departed. They kept this up for three years with no explanation. I knew Indians well enough to know that when the time came they would talk.

One evening while I was feeding the chickens they came. The one who had done all of the talking came over to where I was and said, "I am not an Indian but have been a missionary to them down at Indianola for nearly 40 years. My name is Mormon V. Selman and I want to present to you an English-Ute dictionary that I have had printed."

I took it and thanked him for it and when he saw I was apparently pleased with it said, "I have something here I wish to show you." He took from his jumper pocket something wrapped in a large bandana handkerchief. He was smiling, something I had never seen him do before, as he unwrapped it.

Rich Gold Ore

He brought forth a piece of gold-bearing rock about two inches long, one inch wide and one half inch thick. This he handed to me. I looked at it and it appeared to be some sort of a honeycombed rock with the wax part all strings of gold and some of the cells full of what appeared to me to be pure gold. I handed it back and then he began rewrapping it up.

"It appeared to be some sort of a honeycombed rock with the wax part all strings of gold."

When it was again safely in his pocket he said: "For many years we have sat around the council fires and listened to the old chiefs or the medicine man, tell of the legends and traditions of the Indians of the Uinta Basin.

"One night each year we gather around the council fire and each leader, or underchief, tells of some heroic deeds certain braves have accomplished. When they have all finished then either the medicine man or the highest chief selects the one, or ones, to be rewarded. Usually a small piece of gold is given for the brave to carry in his medicine bag."

Legend about Whites

"A few nights ago as we sat around the fire the tales were told and imagine my surprise, for my name had never once been mentioned, when the chief arose and presented me with this piece of gold. Why that is the largest piece I have ever seen given any warrior and to think he should give it to me is almost beyond belief."

There was a long silence and then he continued: "There is a place out in the basin that the Indians consider holy. No one but the medicine man or the chief dare go there. They call it 'Carre-Shin-Ob,' which means 'There, the Great Spirit.'

"One of their legends has to do with the first white men who went into that country. They claim it was many, many years ago, when seven white men riding horses went there. They were hunting for 'money rock,' which is gold.

They Tortured Him

"They came from the east and crossed Green river at the old Indian ford and camped near Vernal. Then one Indian happened to take out the contents of his medicine bag and in it was a gold nugget. Immediately they asked where he got it and where it came from. He would not tell, could not, for he did not know. They tortured him and he died without saying one word.

"Those white men had sticks that shot fire, which were the first ones those Indians had ever seen and to demonstrate their power they shot another brave. Then the Indians supposed they were some of the Great White Spirits they had been taught would come to them. When the men demanded their bows, arrows and spears they readily gave them up.

Promised Two Baskets

"Then they took a young son of the chief and tortured him until he died, but the chief would not talk. Another son they killed and then the Medicine man told those men if they would go away he would fill two baskets full of that rock they wanted. They picked out two of the largest baskets the Indians had but the medicine man said he could not carry so much. Then they loaned him a pack mule and he departed late in the evening just as it started to rain.

"When he returned early the next morning it was still raining. The white men looked at the gold and were very excited. Immediately two men saddled two horses and took two pack animals and started to back-track the tracks of the medicine man and the mule. In vain the medicine man protested.

"All day long that medicine man was busy making strong medicine that would be stronger than the white men had, even stronger than their fire sticks. Evening came and the two men returned their mules staggering under heavy loads."

He Denounced Them

The medicine man met them and denounced them and called upon his Great Spirit to come to their assistance, for now he knew these were not Great Spirits for if they had been they would have known where Carre-Shin-Ob was. One of the men grew angry with him and tried to shoot him. Another tried but his gun would not fire. You see, their powder was wet.

"Seeing this, the medicine man called upon his people, from the high chief to the little boy to join with him and drive those men away. Grasping sticks, stones and everything they could find, they soon overpowered the white men, and when they recovered their bows and arrows, they made the white men put saddles on their horses and depart, but they went along to the ford to be sure they would not return. Not one piece of their precious or sacred rock did the white men take away with them. For many nights the medicine man was busy carrying all that 'money rock' back to Carre-Shin-Ob."

One of the Indians called Selman said, "Supper is ready. The next time I come I will tell you more."

Waited His Return

Anxiously I waited his return the following summer. Selman had aged considerably. He seemed more pleased than ever to see me. Again we sat on those granary steps but he had little to say. Then it was he told me there were only two men remaining alive who knew the location of their sacred place and on this visit he was going to try to get them to tell their information to some of the younger chiefs or to him.

Two weeks later they returned. He seemed terribly down-hearted.

Late that evening he called me out of my house and said: "As we sat around the council fire I told those two old chiefs how they had honored me by giving me that piece of gold the year before, and now I had a request to make of them. I wanted them to tell someone where that gold was situated so when they died the secret would not die with them.

Left Council Fire

"Those two chiefs left the council fire and went out into the open for quite some time. When they returned, one said, better the secret die with them, than they tell the younger chiefs, for they did not trust them and were afraid they would get the gold and sell it to the white men for whisky, and if they did then the place would no longer be fit for the Great Spirit to dwell in when he came.

> **"Better the secret die with them, than they tell the younger chiefs, for they did not trust them."**

"Then I learned something else from them I had never heard before. One summer the whole tribe went up north someplace for a sort of a reunion and when they returned the medicine man discovered someone had been to that sacred place, took all of the loose gold and had dug a tunnel into the mountain and took much more of the gold.

Went After Thieves

"The old chief who told it said at that time he was a young brave and he said it happened shortly after the Mormons came to Utah. He and six others were sent after the thieves, who left a plain trail to follow. Over the mountains they went and finally down the Provo River, and then south to somewhere near Nephi, where they caught up with some Mexicans who had a lot of little mules and some fine riding horses.

"They killed those men and took the string of horses and mules up Salt Creek Canyon and over Sanpete mountain and back to a spot near the mine where the old medicine men took over. Each night he would take a mule with a load of the ore to that sacred place, put the ore back in the mine and then take the mule a way off and kill it so its spirit could not communicate with the other mules. He said when he unloaded a mule, he would put the leather pack bags on top of the ore so as to be a warning to anyone else not to take that gold."

"When we rode into Castle Gate and lifted that payroll, I never had so much fun in my whole life."

He was silent for some time and then he said, "My Father used to tell of a time in the early days when a pack train came down to Provo and camped at his place for a few days to rest up the small pack mules. He said they loaded those animals with a heavy pack load that did not appear to be very large but it was all those mules could carry.

"The men kept an armed guard at their camp and no one was allowed near. He said they stayed a few days and went south and few days later there was a report that some Indians had killed those men down on Chicken Creek (Levan) and had stolen the mules and horses and what ever those animals were loaded with.

"No one suspected that Indians would steal a pack train and everybody decided it was someone who had dressed up as Indians. They laid the blame on Indians who were innocent but after hearing that old chief talk I am led to the belief that the Indians really did the killing."

The following year no Selman returned. The Indians informed me he had died but died happy.

Follow Trail Farther?

Shall we follow the trail farther? Then let us skip over the years until 1945, to a rooming house on First ave. of Salt Lake City.

He was slowly rocking back and forth in an easy chair when we entered the room. His snow white hair bespoke his age—eighty-seven he had previously told me. His wife said, "Joe, a visitor to see you." Evidently he did not hear her.

Then she said in a rather harsh voice, "Joe—" He tottered to his feet. His right hand dropped to his hip. There was a look of do or die on his face as he turned and looked our way. Seeing who it was, he said, "Was just thinking about the time when Butch Cassidy, Elza Lay and I stole the Castle Gate payroll. A body has to be mighty careful even now and I was just thinking what I'd do if some one sneaked up behind me when you came in. Sit down, son, sit down. What's on your mind?"

"Well last time you told me some day you would tell me about a gold mine you knew about in the Uinta Basin but right now you have me wondering if really you ever rode with Butch. Did you?"

About That Gold Mine

"Sure did, and those were the happy days. Say when we rode into Castle Gate and lifted that payroll I never had so much fun in so short a time in my whole life. That was a good haul. But about that gold mine—

"You see it was this way. Shortly after we divided that Castle Gate money, me and Butch were riding from the Roost (Robber's Roost) out to Brown's Hole and not wanting to be too conspicuous, we were trying to avoid seeing too many people. We cut across west of Vernal and when we entered the mountains to the northwest of that town it began to sprinkle.

"Butch spurred his horse and we rode up a side canyon and stopped near some thick brush. We left our horses here and ran like [heck]. I followed right behind him and into an old mine tunnel we went just as the rain came down in buckets full.

Old Spanish Mine

"Now that tunnel was not over five feet high and three feet wide and was piled about two feet deep with rocks. When we had sat down on some old leather pack bags Butch told me this mine was the one the old Spaniards had worked. They called it 'The Madre de Oro del Uinta' which means 'The Gold of the Uinta.'

"Butch sure knew his history and knew a lot the Indians had told him for he was friendly with them. As we sat there I reached under that leather bag and got a rock. Gosh it was heavy. I took it to the mouth of the tunnel and say it shined like almost solid gold. Butch called me to come back and told me any one who took any of that gold would have the curse of God placed upon him. Said in early days Brigham Young was good to the Indians and they gave him a lot of this gold and one day an old Indian told him if he would go out there he would give him all a pack horse could carry away.

Never Got To Mine

"Brigham told that Indian he could not get away but would send Tom Rhodes with a pack horse loaded with food and clothes if the Indians would fill the pack bags full of that gold. Old Tom, Butch says, made three or four trips out there but never got to the mine. The Indians would not let him go where they got the rock. They said it was the place for their Gods to live.

"I slipped a small piece of that gold in my pocket but when we stepped outside Butch drew his gun and told me to put it back. How the [heck] did he know I had it? But Butch he knew everything. Well I went back and put that piece of rock about the size of my hand on top of one of those leather bags and came back out for I knew Butch was not fooling.

"When it stopped raining, we rode away and Butch made me promise I would never go there again.

Not Afraid of Anything

"When the Wild Bunch broke up, that was what we were called who rode with Butch, it was mighty slim picking for some of us, so one day I got two of the men who I knew were not afraid of God, man or devil or the curse of God and told them about that gold.

"We got eight fine horses, and a pack outfit for heavy work and a light one. Then we rode out and camped about half a mile from the mine for almost a week just to see if there were any Utes watching. We saw none. Then one night I took one of those men and had a hard time finding it for you see the timbers had rotted and it had caved in at the entrance. We had to crawl in on our hands and knees.

"Once inside we lit some matches and there on top of that leather bag was the same piece of gold I had put there years ago. Then we went back to camp and laid these plans. The next morning I was to take five horses and the heavy pack and go down to the ford at Green River and wait. They were to wait until night and go to the mine and get not over one hundred pounds of that gold ore and make a swift ride to where I was, change horses and head for Denver and sell the ore and give me my share in cash. If we made it, then we would make other trips.

"The Indians would not let him go where they got the rock. They said it was the place for their Gods to live."

"I got to the ford and staked my horses around and built a fire and got supper but did not go to bed for I knew they could make it down there by about three in the morning so I sat up and kept the fire going so they would not miss me.

"Daylight came and no men, then sun up and along came three Indians. They stopped and talked with me and just then we saw those two men coming but mighty slow. When they saw the Indians with me they let go of the rope on the pack horse and lit out for the river, jumped their horses off just below the ford and started across. They slipped off the horses and each grabbed his horse's tail and they started to swim across.

"When the horses were about a quarter of the way across they decided to turn back. The men let go and tried to get hold of the bridle reins but both went down and we never saw them come up.

Those Darned Fools

"The horses came out right where we were and say, those durned fools had about one hundred pounds of gold tied in a sack on each saddle. I'll bet they had at least fifty pounds in their pockets and that was why they could not swim.

"The Indians looked at the sacks and opened one and saw it was gold. Then they went over to the pack horse who was too tired to even try to stand up and say, they must have had at least two hundred and fifty pounds of gold on that little pack saddle.

"Now the Indians talked a short time and one got on his horse and rode like the clatter wheels of [heck]. I thought it was time for me to get going but those two Utes told me to stay there.

I Lied Out of It

"About sundown that Indian and an old chief and two or three more came and examined the gold and asked me a lot of questions but I lied out of it. Then they searched me and for once I was lucky for I never had one bit of that gold on me.

"Just at nightfall they reloaded those three horses and the old chief started back alone. The others stayed with me and the next morning let me go to Brown's hole.

"Now I'm telling you if them two men had not made [darned] hogs out of themselves, maybe we might have gotten away with that gold, and if we had then we would have taken more and now I would not be where I am with scarcely a thin dime to my name but I guess Butch was right. He said there was a curse upon anyone who took that gold and as for me, well I'll hold on to that thin dime as long as I can but never again will I go back to where there must be all of a million dollars in almost pure gold just piled up in an old tunnel."

Hundreds of Claims at Deadman Lake

Author's Note

This saga probably continues with other entries after the August 16, 1894, entry. I haven't pursued that thread past this point but I've given you enough information to follow it at your nearest public library. Many libraries have copies of old newspapers on microfilm. You may even locate sources that haven't occurred to me. In pursuing this story yourself, you may feel the excitement I have felt while doing this research. Good luck.

The Vernal Express

Enoch Davis Gold Ore

June 14, 1894

(Reprinted in part June 14, 1994)

Late yesterday afternoon Assayer Kroupa delivered to Lorenzo Hatch his certificate of the value of ore taken from the long lost Enoch Davis gold ore dike in Uintah County, news of the finding was given exclusively in these columns yesterday. The certificate showed that the rock is worth $366.26 per ton in gold—it contains 17.72 ounces in gold as well as 12.38 ounces, in silver, some mercury, copper and cinnabar gold.

Another specimen of twenty-five pounds is to be assayed this evening or tomorrow that is more of an average specimen of the ore in the vein, and it is confidently expected that this will go even better than did the smaller piece of yesterday.

There is little or no doubt but that the entire dike is wholly outside of the boundary line of the Indian reservation. All three of the locations are in the Uintah County Mining district, and their locations are recorded in the books of said district and certified to by B.O. Colton, recorder of that district.

M.M. Warner of this city has been taken care of by the finders: He has been given a slice of the claim, and is an owner in the big bonanza, which is beyond question the richest mine in all of rich Utah.

Found at Last: The Fabulously Rich Enoch Davis Gold Mine by Dallin and Hatch

The long lost, and eagerly looked for Enoch Davis gold mine has been found. George Dallin, a prospector from Salt Lake, and Lorenzo Hatch, a prominent citizen of Vernal, are the lucky finders. Both of these gentlemen are in the city, and they, more than anybody else, will be surprised to read this announcement in the Dispatch this evening, for they have used every endeavor to keep their secret.

Among certain prominent members of the mining fraternity of the territory for the past eight months or more, there has been much comment and speculative talk indulged in concerning this particular gold bearing ledge or vein. Its exact location has been known only to Enoch Davis and a very few others. It has been reported as being very rich, so rich that one can cut gold out of the ore with an ax, says Davis, but Assayer Kroupa of this city who is now hard at work making an assay of a sample of the ore expects it to run only about $300.00 to the ton, which is rich enough in all conscience, especially considering the fact that the vein is a 15-foot vein. It is located in the mountains east of the Uintah Reservation, some say in the reservation, but Messrs, Dallin and Hatch assert positively that the mine is wholly outside the boundary line.

The history of and interesting incidents concerning the finding of this claim are these: Enoch Davis who is now awaiting the death sentence in Salt Lake penitentiary for the killing of his wife in Vernal two years ago, his case having been carried up to the supreme court of the United States and passed upon by that body, while Cass Hite, well known as a frontiersman and miner, was serving out his sentence for the killing of Kohler, but who has since been pardoned, told Hite of his discovery. Davis believed his find to be located within the boundary of the Uintah Indian reservation and understood that he would have to wait for the reservation to be thrown open before

"The party is looking in vain for the lost treasure, now securely in the possession of others."

he could obtain title to the claim and work the same. Feeling that he was to lose his life Davis drew a diagram of the country and explained to Hite how and where to go and find the mine, saying that when he found it he should give one-fourth of it to his (Davis') children, one-fourth to M.M. Warner of Provo, the attorney who defended Davis so valiantly without compensation or hope of reward, and the balance, one half, Hite was to have as his own.

Hite, on getting out of the penitentiary and after attending to some urgent matters of business, at once set about devising means of exploring the country and finding the hidden treasure, the 15-foot vein of fabulously rich gold ore. He had spent much money in conducting his trials and was not well fixed financially as he once had been. He interested other parties, including prominent public men in Salt Lake, and a purse for an exploring party was raised. That party is now out in the vicinity of the mine on its third trip wallowing through the snow and looking in vain for the lost treasure, now securely in the possession of others.

In some unaccountable way Cass Hite's secret has become known to several others, and even now there are many men in the vicinity of this find, and near Hite's party watching its progress and prospecting in the hills near about. Davis had a partner by the name of Hatch in this find, a brother to Lorenzo Hatch mentioned herein. Lorenzo had been to the find with his brother. He was expecting to go out and locate it this summer feeling confident himself, that it was outside the boundary line of the reservation. Learning that the Hite party was looking for it, made him go to the mine earlier, and only four days ago it was located. Mr. Hatch, when near the vein, stumbled across Mr. Dallin looking across the valley through a field glass. The two went direct to the vein and located it. They are here now with samples of ore from it, having the same assayed.

Near the point of location Messrs. Dallin and Hatch discovered some digging. Knowing that Indians never dig, this seemed strange indeed to the men and they began to look around. Their search was profitable. They found the decayed remains of a white man's camp, including scraps of blankets, pieces of cooking utensils, etc., and nearby the charred remains of some human bones. Evidently a man had been killed and his body thrown on a fire and burned. It is now recalled that some six years ago a Frenchman came into Vernal and went into the Indian reservation, ostensibly to sell beads and other trinkets to the Indians, and that he was never seen again. Undoubtedly the Frenchman was killed and robbed by the Indians and these are his remains.

"There is no doubt of Uintah County becoming one of the best mining counties in Utah, if all reports are true."

June 21, 1894

The great gold find near Gilbert Peak is creating the greatest excitement of any find that has been made in the west for years. There is no doubt of Uintah County becoming one of the best mining counties in Utah, if all reports are true.

Lorenzo Hatch and Mr. Haladay, were in Vernal this week and were as full of business as a cocoanut is full of milk, but would not say much about what they were doing, or what they were going to do. They recorded some new claims, however, and those who have talked to them, say that they are strictly in it.

The Cass Hite party are keeping the secret of their doings to themselves, but a party who visited their camp picked up a piece of rock among some samples that they had in camp, and could see the gold in nuggets sticking in the rock, showing that Mr. Hite was on to something better than was first found, and will surprise the mining fraternity when he makes known to them and the community what he has found.

Vernal is nearly deserted this week, and parties are leaving daily for the new camps, and next week the Express may be able to give a more complete and authentic account of the camp and what its future prospects are. Every one seems satisfied however that it will prove to be a genuine bonanza and a bright prospect is predicted for the people living in this valley.

June 28, 1894

Mr. Holiday, one of the parties interested in the gold find, called at our sanctum last week and gave us a few pointers in regard to the discovery. Their claims are on the headwaters of the east and west forks of Ashley and the east fork of White rock creek. The dike runs nearly due east and west and is fifteen feet wide and is composed of gneiss, porphyry and quartz. The dike cuts the formation and dips lightly to the north and has every appearance of a true fissure vein. They have bought a mill site of some parties owning a claim adjoining them and when a mill is built can dump the ore from the mouth of the tunnel into the mill.

Mr. Holiday says that it is all a mistake about this being the Enoch Davis mine. They found the bones of two men and three horses with some camp material. The camp showed evidence of a hard fight and the men being murdered and the victims of the crime being burned to destroy any evidence of who and what they were. It is a well known fact among old settlers of this county that there are good mines on and near the Uintah Reservation and they are jealously guarded by the Indians.

In early days a man named Rhodes, would go into the Uinta Mountains and come back loaded with gold, but he would not give his secret away and no one knew where he got it. On his deathbed he told his son where it was, and afterward the son was killed by the Indians while after the treasure, and thus the secret of its whereabouts was lost. Several parties have been organized at different times to hunt for the lost mine, and some of them never returned.

Probably the camp found by the Hatch party was one of the camps of one of these parties.

There is no doubt that when the Enoch Davis mine is found, that it will be free gold and very rich. With a few hundred men hunting diligently, there is a very small chance for the secret of its whereabouts to remain a secret much longer, and when it is found it is the supposition that the Rhodes mine will be found.

The Hillsdale Town site Company has been incorporated for $100,000 by S.M. Brown, W.C. Britt, S.D. Colton, Ed. F. Harmston, and E.W. Davis.

They have made a survey of the town and have plotted forty blocks of eight lots each, 75x150 in size, and have filed the plot with the recorder. The embryo city will be called Hillsdale, and is said to be situated in a beautiful location near several small lakes, and will always have a bountiful supply of pure water.

"There is no doubt that when the Enoch Davis mine is found, that it will be free gold and very rich."

The people of this valley want to get a move on them and see about a road to the new mines. It is reported by good authority that there is only about eight miles of road to build, there being a road already built to within that distance of the mines. This is no idle chatter, but it means a good deal to the farmers here, as well as the business men. The U.P.R.R. is said to have already offered to build a road to their nearest railroad point provided the mine owners will bring their freight that way. A good many of the claims are owned by our citizens and they will naturally favor their neighbors if they will only make an effort toward helping themselves by making this road so that supplies can be had from this valley and outside freight come this way. Do not procrastinate, but see to it at once.

July 5, 1894

As you wished an article about the new gold mines, I will try and tell you about what I saw while there. I have seen a great portion of the dike or vein for about twelve miles or more. It is the biggest thing I ever saw, a true fissure vein, or rather two fissures, about 800 feet apart, which crops out in several places. Our party is located on the south fork of Ashley, the vein showing up for over sixteen feet in width.

There is a fine chance to make a wagon road from the Carroll mill, up the south fork. The people of the valley should look after this matter immediately, as it could be done with very little labor.—W.T.

July 26, 1894

"Why don't you boom the mines?" That is the question asked of the Express almost every day. The Express has given its readers all there is to give in regards to them, and that is there are plenty of prospectors in the hills and very high assays from the prospect first discovered. Good mines close by are what the people of this valley need in their business and the Express will boom them plenty when such mines are opened up. Hundreds of claims are staked off. A few are being worked by their owners for the purpose of finding out their value. The majority is waiting on these few thinking that if the claims that are worked are good, theirs will be, and on the other hand if they are not good, they will not be out any thing for labor etc., and these are the very ones who are so anxious about having the mines boomed.

Several bona-fide prospectors are waiting anxiously for returns from samples of ore sent out by them, and if the assays already had are genuine theirs will be just as good or better.

The work on the Deadman mine, has been stopped for a short time and M.M. Warner, goes to Salt Lake this week on business connected with the mine.

Frank P. Warner, brother to M.M. Warner, called at the office of the Express last week and says that it is his opinion that the mines will prove to be a genuine bonanza. He found a ledge bearing free gold, about two miles west of Deadman, and has sent samples of the rock to Salt Lake to have them assayed.

Assays recently made on the samples brought back from the Uintah County gold discoveries by the prospectors sent out by Major Stanton, Bob Conner, and others, yielded encouraging returns. The gold is found in a red talc formation, and appears to be in streaks. Some samples carried as high as $1700 in gold to the ton, while other lots that looked similar carried none of the yellow metal. This is a peculiarity as yet unexplained, but it illustrates the old saying that "Gold is where you find it." The samples came from the region bordering the Indian Reservation on the east.

Aug. 9, 1894

Enoch Davis, the wife murderer, whose days on earth are rapidly drawing to an end, is held in solitary confinement, the only person permitted to communicate with him being the guards who preside over the station adjoining his cell.

The prisoner, as has been his conduct from the very first, is very retiring, shrinks before the gaze of the visitor, and keeps his reflections practically to himself.

For weeks he has cherished a hope that the expedition upon which he sent Cass Hite in search of a lost mine of fabulous riches, would develop something to check the course of the law's decree. His first inquiry in the morning, his last at night, is for mail, but none, save an occasional letter from his lawyer, has reached him. He states that Hite promised to write him concerning the search, and his failure to do so has made the doomed man very severe on him. Davis feels confident that if Hite kept in the trail marked out by him upon maps that were improvised in his cell and jealously guarded until the latter's release, that the mine, which lies on a line due east, some eighty miles from Salt Lake, has been recovered. The more reasonable conjecture would be that it had not been discovered, although no course of reasoning can convince him of it.

Davis's story to Hite was either an unparalleled marvel of fact or an example of fiction that few have achieved. Not a link in it was missing, and when Hite left it was with maps upon which he expected to go straight to the treasure, which, the prisoner says, is in one of the recesses of the Uintah Reservation.

"Convicted of murder and sentenced to death, he began to cast about for some means to cheat the executioner."

Davis claims to have discovered the whereabouts of the mysterious bonanza years ago, but, for reasons as mysterious as the silence of Hite is to him, maintained absolute silence until, convicted of murder and sentenced to death, he began to cast about for some means to cheat the executioner. At the same time Cass Hite was doing time for having killed a man, and Davis awoke one morning to learn that Hite was his cellmate. It was not long until he found himself an eager listener to Hite's tales of adventure upon the riotous cataracts of the Colorado River, the canyons of which he has explored as no other, and Davis himself soon began to talk. He told Hite of how, in boyhood days, he had been the playmate of "Giant" Rhode's sons, and how he had seen the Hercules, whom he regarded in childish awe, issue from the mountains bearing with him nuggets of gold that soon filled a wooden bowl; how all efforts to hunt out the mysterious source of these riches had been baffled by the wary old giant, who had withheld it from his sons until they acquired years when they might appreciate the value of his secret.

Hite was electrified, and urged the doomed man on. Davis told him how finally he left Rhodes plateau and went to a fort in the Uintah

Reservation, where he began his apprenticeship at the anvil. Several years went by, and Davis, finally a graduate of the forge, was made smithy at the fort.

One sultry afternoon, said Davis, the sweat pouring from his brow, while he stood puffing on his pipe without the shop, a traveler rode up, and, dismounting, asked that his animal be shod. He had seen the face before, and as he fitted the iron and drove the nails it came back to him that his unexpected customer was none other than one of the sons of "Giant" Rhodes, with whom he had played in youth. Davis revealed his suspicions to the traveler and work was temporarily suspended while they sought the fort canteen to celebrate the reunion. "Giant" Rhodes, said his son, had succumbed to the awful embrace of a giant cinnamon bear, but not before the boys had been shown the trail to the mysterious mine.

Davis says he seized the hour of golden opportunity, and, with his old playmate, went to the mysterious mine. The doomed man swears he saw the riches, of which he had often dreamed, in all their matchless abundance, and, withdrawing with young Rhodes, returned to the fort, where he renewed his labors at the anvil. Rhodes continued his journey, said Davis, and a few months later followed his brother to the grave.

Thus was all knowledge of the mysterious mine obliterated, save that in possession of Davis.

"Verify this," exclaimed Hite, "and you are a free man."

Davis seized the straw that offered, made the maps, and with them Hite, who had meanwhile been pardoned, started on his journey in search the lost bonanza. Davis relies upon its discovery as the only thing that will save him from the sentence that hovers over him, and, having heard nothing from Hite, is almost a madman at times.

With the proceeds from the mine he feels convinced that there is every show for outwitting the executioner, whose dreaded voice he listens for by day and night.

Aug 16, 1894

Tom Pancake and John Regan returned Monday evening from an extended trip south of here. They were on a pleasure excursion and incidentally did a little prospecting. They were close to the Gilbert Peak mining section, but could not get over the mountains on account of having no pack animals. The reports they heard were encouraging concerning the new camp, and the chances for it becoming a big gold producer are very bright. One sale is recorded wherein a claim owner sold a forth interest in his location for $12,000 to Salt Lake men.

The buyers agreed to put the mine in shape, buy a mill and in fact develop it to a paying property, in addition to the purchase price, and that, too for a fourth interest only. They also offered the owners $130,000 for the other three fourths, but the offer was refused.

Although only a small amount of work has been done, the lead shows up splendid. The road from Green River to this camp could be easily put in good shape, and all of that trade could be brought here and should be.

There are about five hundred men in the camp at present and hundreds are on the road. Tom and Jack report a pleasant trip and a good time.

The chief interest in mining circles is now centered on the rich mineral fields of the Uintah and Uncompahgre Reservations, which are to be thrown open to the public, and any information concerning this region is received with interest. The following extracts are made from a letter received by a well-known Salt Lake mining operator from a party of prospectors that recently sent out to the Uinta Mountains, north of the reservation. The letter was written at Fort Duchesne:

"The mining excitement is high here. Parties are pushing in from Denver, Cheyenne, Henry Mountains and Colorado River region. The camp at Deadman is the center of attraction, and forty miles north of Fort Duchesne, is about as tough a place as they make them. Bad men and worse men from Creede, Denver, Cheyenne and Helena are there."

"They have ore which, it is said, runs from $280 to $1500 in gold. Cass Hite tells me that he is camped twenty-one miles northwest of our place. He has located some quartz claims there, but is not bothering with placers yet. His belief is that the field is a rich one. Last night he showed me a piece of rock in which I could see the gold with my naked eye. What he has isn't a fake, and it came from the head of the stream which washes down our gulch. I will ship you the samples we have at the first opportunity. A party of Denver men is here on their way to a point west of Deadman."

"I have seen a magnificent stretch of country right in the mountains and east of Lake Fork, about thirty to forty square miles in area. It is watered by a single mountain stream and four or five lake streams. The soil is the richest I ever saw in Utah and the grass and wild grain for twenty-five miles grows as high as a horse's back. People in Salt Lake have no idea what a country this is. The mountain range and Bad Lands shut off the best country from ordinary travel, and unless a person travels by the Indian trails, as we did, he will not see the fine places."

Precious Metals at Dyer Mine

The Vernal Express

The Story of the Historic Dyer Mine

December 10, 1926

William Shaefermeyer

Many stories of mysterious finds of rich gold and silver-bearing rock in the Uinta Mountains north and northwest of Vernal have been told so often that they have become part of the annals of the Uinta Basin. Some of these tales, perhaps, had no foundation of truth whatever, but with absorbing interest has the writer listened to them, and this interest led to the resolve to prospect parts of this mountain region.

Many years prior to the opening of the Indian reservation to settlement, the writer heard the first of these tales of riches found and lost. It was said that a fabulously rich gold mine existed far up in Whiterocks canyon. This deposit of wealth containing ore was known as the "Indian mine" and according to the story only one white man knew of its existence. It was said that the location was closely and jealously guarded by the Indians and that the red men had issued warnings to the whites to refrain from seeking it. No one knows what became of the solitary white man knowing of this mine. It is known, however, that in more recent years a white man offered the Indians something like $40,000 for the secret of the mine's location, with the understanding that the transfer of the sum mentioned, if accepted by the Indians, would give the white man ownership of the treasure. According to the story, the white man was confident that in a short time he could garner sufficient gold from the mine to insure him a profit on the transaction. The offer was declined by the older Indians, but the younger red men were disposed to accept the offer.

The writer also was told that in the region where the "Indian mine" was located a deer hunter stopped at a little mountain brook to allay his thirst. Stopping over a pool he saw on the bottom of the stream shining particles of yellow metal and when he examined a handful of gravel he had scooped from the bottom of the pool he discovered several nuggets of gold. Then a shadow passed over the pool and gazing upward the hunter beheld an Indian with a gun pointed his way. In threatening tones the Indian told the hunter to "vamose" but first to throw back into the water the gold he had taken from it. The order was obeyed, and then the Indian escorted the white man from the region over devious trails. Thus the gold and its riches were lost from the white man's ken for all time, as the hunter said to have later stated that he had not the faintest idea of its location.

Other stories tell of a white range rider hunting cattle on Mount Dyer picking up some rich gold float, and of another man who was hunting in the breaks of a high ridge between Pot Creek and McKee draw on Diamond mountain, stumbling on an outcropping of rich copper ore. Neither of the locations has ever been found again.

"He did not realize the richness or the importance of his find and, perhaps, was careless."

Another story tells that many years ago a man found a place where he dug out a bright white metal that coated the blade of his shovel as if it were plated with silver. This metal, apparently, was gallium, a rare and mysterious metal found in but three or four locations in the entire world. This metal indicates a deep source of mineralization and is usually associated with gold and zinc.

Many other rich finds in the Uinta Mountains are recorded in the stories of this region, and surely these tales have some sort of a foundation of truth. We know that in the Uinta Mountain range there are places where lead ores predominate, other places where copper predominates, others leading in zinc and silver and iron ores. It is not at all unreasonable to assume that gold may be found in this region, and the "lay of the land" indicates that if gold is found the wealth of the deposits may be far beyond the wildest dreams of prospectors. To some people such an assumption may appear as merely an idle thought, but to the real prospector who understands his business it means a great deal. The first thing he does is to study the geological formations of the region and from these he obtains the information as to whether or not the conditions are right for mineral deposits. The outcrop of surface ores and the geological character of the country rock with which these ores are associated are the elements giving a true index of what may be expected farther down, for the old adage "Poor float, poor mine, rich float, rich mine" can always be depended upon. A good prospector never goes forth looking for a "lost" mine. His business is to make a close scientific study of a region, and he soon learns whether or not it will pay to open up a strata.

So it was with the writer; first he examined the geological conditions, and soon he found a highly mineralized belt following a strong fracture zone in the Mississippian limestone and the Ordovician quartzite. These are the most noted ore-bearing formations in the world, and it was found that, without exception, the stories of the rich Uinta Mountain region finds were centered along this fracture zone.

History of the Dyer Mine

Now comes the story of the old Dyer mine, and a mine it was, indeed. Some of the richest copper ore in the world (exclusive of native copper) was taken from this mine. Out of a hole in the ground 50 feet square and 75 feet deep was taken copper to the value of $700,000. This proves without a shadow of doubt the richness and importance of this mineral-bearing belt. Many of the incidents, the controversies, connected with the locating and mining of this ore body will perhaps never be known, for they are shrouded in the misty fog of romance.

The first discovery of the Dyer mine was made by Ira Burton in 1884. Mr. Burton and Andy Holfan were camped on Diamond mountain, looking for cattle. They had made camp at what is now Bullionville near Big Brush creek, while Mr. Burton was hunting near what is now Dyer Peak. On the side of this peak he stumbled across rich copper ore scattered promiscuously over the mountain side. He gathered a bunch of the ore and returned to the camp on Diamond mountain. He did not realize the richness or the importance of his find and, perhaps, was careless. A short time after he arrived at his camp a wandering prospector, L. P. Dyer, appeared and he was shown the copper ore picked up on the mountain side.

At once the prospector saw the importance of the find. "Boys," he said, "I will make you rich if you will show me the place where you got this stuff." Mr. Burton pointed out the peak in the distance and said the place was near the top.

Dyer lost no time in departing for Vernal, and outfitting with tools and provisions he traveled by another route to the place indicated by Mr. Burton. He located the entire outcrop in his own name, refusing to give Mr. Burton and his companion any interest in the claim.

The full tonnage of ore taken out of the hole by Dyer is not known. The ore was hauled to Rock Springs, Wyo., and yielded an average assay of 49.47 percent copper, 26 ounces per ton silver, and $6 per ton of gold.

In 1899 a 42-inch water jacket blast furnace was installed and ran for two years, producing copper bullion 95 percent pure. A company was formed, known as the Copper Summit Mining company. About 1904, the big pockets being worked out and no systematic development being introduced, the company discontinued operations, and since then the property has been idle. It is now controlled by the Idaho Reduction company. A well known mining expert a few years ago examined the Dyer mine and issued a report to the effect that at depth plenty of good ore is to be found.

Between the years 1891 and 1917 the output of ore in the Carbonate mining district was 4377 tons, and this ore yielded $395,655 in copper, $63,497 in silver, $18,857 in gold and $114 in lead. The district has 25 patented mining claims.

Sky Pilot and the Gold

The Salt Lake Tribune

Gold Mine in the Sky

Nov. 5, 1950

James P. Sharp

The material for this story has been gathered over a period of many years, under various circumstances and from many people who told their stories freely. I have followed up a number of clues and have asked many questions.

Among those who told me the most interesting stories are Mr. Thomas Welsh of Burnt Fork and Harry Somsen of Cokeville, both of Wyoming; John Jones of Tabiona or Hanna, Utah; Arlin Davidson of Meadow, Utah, and Lloyd Henry of Salt Lake.

The stores they told lead to a real, or mythical, lost gold mine high in the Uinta mountains, which is supposed to be somewhere near the head waters of Rock creek in the Uinta basin.

For over 20 years I was livestock appraiser for one of the largest banks in Salt Lake. It seemed to me that whenever there was a bad storm some bank official would get a bright idea in his head and send me out to look after some loan or other. It was in that capacity, about 1928, that one night I rode a very tired horse into Burnt Fork. The snow was deep and it was bitter cold. There was no hotel in the town so one, "Tom" Welsh, took me to his home for the night.

Sat Around the Fire

After supper we sat around a large fireplace while he told stories of men who had made and lost fortunes in the cattle business almost overnight. I was enjoying the heat from that fire as well as his stories.

One I remember him telling concerned a young cowpuncher that used to take a pack horse and go back in the mountains for two weeks every summer. He said this man, he called "Jim," came back from one of those trips and went up north. When he came back he had 2000 head of cattle. He knew Jim didn't gamble, so he looked in the papers for an account of some bank or train being held up but saw none. Then he said, "Where in the [heck] Jim got the money to buy those cattle is a mystery to me."

Then I met Somsen of Cokeville. He told me he used to ride the Sublette range along about 1900 and one day one of the cowboys took a pack horse and rode away somewhere in the mountains, and when he came back he went up into northeastern Wyoming and purchased 2000 head of cattle, which he turned on that range. He said some of the ranchers got mad at him because he did not own a foot of ground, so he took another trip. When he came back he purchased two ranches, both large ones, which he paid for with cash. One was down on Ham's fork, the other up above Cokeville. I asked where he got the money and he said some of the other riders supposed he had found a cache of gold that some outlaw had buried. Somsen said his name was Jim Chrisman.

You see now, we have sort of a tie to begin with.

About 1928 I was hunting deer over near Tabiona, Duchesne county. It was bad weather and worse hunting. However, I had just shot a deer when a man who said his name was John Jones rode up. We talked about the storm and things in general. He told me that a recent storm had spoiled his plans. He said he had heard of a lost Spanish mine up in the Uintas and was searching for it when the storm came. Here is his story. "I was away up at the head of Rock creek and had just found an old cabin, some very old pack saddles, an old shoe and a few other things when the storm came. I could see a good trail leading up the mountain from the cabin and I knew that would lead to the mine. Well, I stayed in the cabin that night and say, there was some mighty rich gold ore strung around in it. Next morning the storm drove me out, but you just wait until next summer and I'll be sitting pretty."

"Do you think that was the Rhodes mine?" I asked.

"You mean the one Tom used to get the gold from which he gave to Brigham Young?"

"That's the one."

"Naw, 'taint up there. His mine is lower down. Say man, that mine is up so high a wild goose has to get a running start to fly over it. That ain't where Tom used to go."

The Eastern Stranger

Now just a few words before I start on the combined story. I cannot give the names of all that have spoken to me about this particular mine. If I did, it would be like listing the telephone directory.

According to the information I have, sometime between the end of World War I and 1930 (people do not agree on dates), a man rode into Bridger valley. Apparently he was a well educated eastern man. He stayed at one of the ranches for a few days and then one of the cowboys joined him and they made a camp of their own. A few days later another cowboy joined them as their cook.

The first one to ride into the valley we will call "Sky Pilot," because he was, at times, very religious. However, he was also a card shark, a sort of hypnotist and a clairvoyant. The first cowboy to quit we will call "Jesse James" because he was quick on the trigger, and the other we will call "Mulligan," for he was always cooking Mulligan stews.

"He said he had heard of a lost Spanish mine up in the Uintas and was searching for it when the storm came."

A few days later what appeared to be two parties, consisting of nine men, rode into the valley (one said seven, another 11, and one nine). It has been hinted that most of those men were the remnant of the "Wild Bunch" of Butch Cassidy days. Be that as it may, they were all well armed and expert shots. They had two pack horses to carry their food and camp equipment. They made camp with Sky Pilot and the other two men. Supplies were purchased and paid for by Sky Pilot, who seemed to have plenty of ready cash.

Refused to Go

The following day a gentleman and a lady drove to one of the ranches and stopped. It was said that he was an assayer, or chemist, who could tell if a piece of ore had gold in it by placing some kind of acid on the rock. One man said they came from Rock Springs while another said Salt Lake was their home. We will call this man the assayer.

Next morning the men all rode up to the ranch. They had their horses packed and an extra horse which was for the assayer. He refused to go one step with the outfit. All of the men returned to their previous camp except Pilot and Jesse James.

Sky Pilot told the assayer and his wife there was no danger and said he could prove it. He produced a new deck of cards that had never been opened. He had the assayer break the seal and while he (the assayer) was shuffling the cards, he told them that spades were bad luck cards, clubs meant trouble, hearts were love and affection while diamonds were riches.

Now he had the assayer's wife cut the deck and draw one card. It was the king of hearts and represented her husband. They were again shuffled. During all this time, and that which followed, Sky Pilot never once touched the cards. The assayer drew a card—the queen of hearts, which was his wife. Then the cards were shuffled, cut and shuffled and finally the assayer's wife was told to draw cards from any place in the deck and leave them face down. She did and then turned them over and had the ace, king, queen, jack and 10 of diamonds—riches untold.

Left His Pistol

The assayer decided something was wrong so he faced the rest of the cards and found it to be a regular deck. His wife decided he would go. He consented, on one condition—that he leave his pistol with his wife and go unarmed.

The following morning after they called for the assayer, Sky Pilot insisted that he be blindfolded. Then taking paper and a pencil, he drew a rough map and said as he made an X, "Here we will camp tonight. Just below us there will be a small lake or pond. The stream flowing into it is full of trout. Feed for the horses is on all sides. Just above us is what appears to me to be some ancient castles. The spirit of a departed Indian princess is dictating to me every move and action. Remove the handkerchief."

The men looked at each other as if mystified, for well they knew the country that had been described, even to the huge rocks which resembled castles.

They reached the designated campground some time before sunset and caught plenty of pan-size trout for all. Still in suspense, they waited for the morrow.

Breakfast over and the horses saddled and packed, Sky Pilot asked one of the men to blindfold him again. Again he marked a crooked trail for them to follow which he said was up and over the high mountains. Then he said, "We should reach camp sometime before sunset. We will find an old cabin and plenty of wood for our fires and feed, and water for ourselves and our horses. The spirit of the princess assures me this."

One of the rougher men took the map and led the way, for it was easy to follow, and possibly one hour before sunset they stopped. No cabin was in sight. Sky Pilot said, "Just around the next turn and quite a way up on the left hillside we will find the cabin." It was a very old and dilapidated shack, but was the place they had been led to. A well-defined trail led up the steep, rocky mountain from the cabin. Some were anxious to follow the trail right then but Sky Pilot said they must follow the directions of the spirit and look for the mine on the morrow.

Claimed Half of Gold

After supper all of the men, except the assayer, Jesse James and Mulligan went off by themselves for possibly one hour. When they returned all were silent and they went to bed.

Mulligan had breakfast ready bright and early the following morning. There was no need to call for everyone was raring to go to that mine.

Breakfast eaten, Sky Pilot had the men all gather around him. He told them the spirit of the departed princess had led him there for a purpose. It was that he should have half of all the gold found to organize a church. He was to be prophet for that organization and the money he was to receive was to be used to build churches.

The men looked at each other. Up until that moment it had been fully understood that all the gold found was to be divided equally, share and share alike. They wondered at his actions but did not wonder long, for he called for someone to blindfold him. Then, taking a pencil and paper, he called upon the spirit to direct him to the gold mine. No lines appeared upon the paper. Again he called but still no lines. Then he called upon God to direct him but apparently God was not at home or was busy out in the Garden of Eden (that is as it was told to me) for still that pencil did not move. Then in anger Sky Pilot took his handkerchief from his eyes and boldly announced he would find the mine without the aid of God or the spirit of the princess.

"He instructed the assayer to declare whatever kind of rock the men brought in worthless."

Time Passed Slowly

Now the leader of one of the groups saw that some action had to be taken so he took over. He placed one of the men in charge of the assayer and Jesse James and instructed him not to let those men leave camp that day. He instructed Mulligan to have a good meal for them along about sunset. He paired the others off and took Sky Pilot with him.

Time passed slowly for those three men. Finally they got to talking and the guard seemed to take a liking to both of them. Presently he told them that while they were all away the previous night, Sky Pilot and Nero had made it plain that both of them were to be killed as soon as the mine was found. He instructed the assayer to declare whatever kind of rock the men brought in worthless. In that way he might save his life. He gave him one of his pistols and told him to use it when the time came. He also told them both of their horses had been started along the back track the night before. From one the hobbles were removed but the other was still hobbled. They had decided to kill the horses but knew the shooting would cause alarm so started them back.

Nero Killed Him

Very near sunset Nero and Sky Pilot returned. Then two others came. All were sullen and quiet as they ate their supper. Just as it was getting dark the last two returned.

"What did you two find?" Nero asked.

"Nothing."

"It's a lie. Search them," shrieked Sky Pilot.

Nero stepped forward. The man reached for his gun but was too slow for Nero killed him. His companion turned and started to run away but Nero shot him in the back. They brought him back to the fire and placed him on the ground where Sky Pilot searched his pockets. They were full of very rich gold-bearing rock, some of which was almost pure gold. Nero looked at it and asked, "Where did you find the mine?"

The wounded man was in great pain and was suffering terribly. Finally he said, "We found the mine away over in—aw go to [heck]." This so enraged Nero that he killed him on the spot. Then he ordered the assayer to get his acid, which he did. He took some of the samples and placed his chemicals on a few and then said, as he was holding one in his hand, "They are iron pyrites, sometimes called fool's gold. Worthless."

Drew His Gun

Nero looked at the gold and then at him, doubtingly. Slowly he drew his gun but before he could fire a shot rang out and he fell dead. Who fired that shot will never be known, for he fell as everyone seemed to be shooting at anyone in sight. Jesse James and the assayer were doing their share of the shooting as they backed away from the light of the fire.

In the darkness they both ran along their back trail. All night long they walked or ran and when morning came, caught up with their horses which they caught and mounted. The one piece of gold rock the assayer had in his hand ready to test when the shooting commenced he had slipped in his pocket and it is reported that sample is now in the University of Mines at Laramie, Wyo.

A Closer Look at Caleb Rhoades

The Salt Lake Tribune

His Secret Died with Him

January 21, 1951

Jean Miles Westwood

Here's another piece of the legendary fabric that has been folk woven about the lost gold mine of Uintas.

Here's another piece of the legendary fabric that has been folk-woven about the lost gold mine of the Uintas.

Editor's note: During the past several months The Salt Lake Tribune Magazine has carried several stories about a legendary gold mine in the Uinta mountains. The stories were gathered by James P. Sharp, who has made an intensive study of the lore of the lost mine or mines.

Mr. Sharp wrote his articles from material originating in Wyoming, from the Uinta basin and from Kamas valley. The following story relates yet another legend of the lost mine and originates in Carbon county.

I have been following with interest the series of articles in The Salt Lake Tribune Magazine section on the legends about the gold mine in the Uinta basin. When I read the first article, about the sacred mine of the Indians, it made me wonder if this was the same mine I had heard about. And when I read the second article and Mr. Sharp mentioned Tom Rhodes, I waited eagerly to see if he would tell the story I knew. But the note with his story on the Wyoming version said that this third article was the last of the series. And I still have a story he has not told.

So I would like to add my bits of proof, or a least additional legend, to the story of Utah's fabulous mine.

Copied a Diary

Some five or six years ago I obtained permission from one of the early settlers of Price, Ernest Horsley, to copy his pioneer diary. He had kept an unusually well documented record of the early days of the town, with the idea of someday writing a history of its settlement. Since he had held many offices in both the town government and the church, he had access to practically all the early records and had the figures to back up his stories. At the time I read the diary he was very old and hoped that I would take over the task of writing his history for him.

He is now dead, but members of his family still have the diary, and I believe that parts of it are included in the Carbon County History published by the local chapter of the Daughters of the Utah Pioneers. At any rate, I copied the most essential parts of it. He also gave me detailed biographical sketches of 10 or 12 of the leading pioneers.

One of these has always fascinated me and I think his story adds a little to the legend. The first settler of Price was Caleb Baldwin Rhodes, son of Thomas Rhodes and Elizabeth Foster. He was born in Illinois and when he was 10 years old, in 1846, he came west with his father and others to trap and hunt furs. This group helped rescue the Donner party in the Sierra Nevada mountains. They then returned to Utah and eventually settled in central Utah, living in Lambs canyon, Rhodes valley, Salem and other places. I am positive that Thomas Rhodes is the same Rhodes of the Kamas stories.

Was a Trapper

Tom was a trapper, scout, farmer, prospector and Indian interpreter. From the time he was 10, Caleb lived his father's life. He spent as much time trapping and hunting as possible, with just enough farming to give him a place to call home. Eventually he married, but he never settled down.

In October of 1877 Caleb and a brother-in-law named Abraham Powell came down the Price river trapping and hunting. They built a log cabin close to the river in the southeast corner of what is now the town of Price. They made this their headquarters for the winter and were so impressed by it that two years later, in January of 1879, Caleb and two other men, Frederick E. and Charles Grames, came back to the valley to settle. Abraham Powell had been killed the winter before by a bear. Caleb had told all his close friends back in central Utah about the wonderful isolated valley, and throughout the next years there was an ever increasing stream of new settlers.

Caleb Rhodes was a powerful man, much feared, much respected, and much talked about among his pioneer friends and neighbors. He was tall, large and heavy-set, weighing over 200 pounds. He had a dark complexion and keen eyes. It was said of him that no one ever passed his door hungry and that all children loved him, but no one ever asked him a personal question more than once.

Settled the Argument

If he wanted to tell you something, well and good, but his life was his own and he resented anyone who pried into it. He was hot-tempered. One of the stories Mr. Horsley told me was that Caleb and another man, C. H. Halvorsen, had an argument over the fencing of a town lot by Mr. Halvorsen. Caleb came with his revolver on his hip and an ax in his hand and chopped down the fence. That was the end of the argument.

Caleb settled two and one-half miles northwest of Price in a meadow with a spring of brackish water which no one else could drink but which he said he liked. He helped to build the first irrigation ditches and cultivate the first land. He was always on hand to help a new settler dig a dugout along the river bank and get settled. But best of all he liked to hunt and trap, so no one paid much attention as he came and left. He would be gone for 10 days or two weeks and then stride in and give each family deer meat or pheasant or bear. Everyone agreed that he was honest and upright but "close to nature," mainly silent, preferring action to words and very insistent upon his rights.

Helped His Friends

Some of the other settlers have told me that Caleb had conceived the idea of building a settlement in this remote valley which would consist of the people he liked best and live the kind of life he liked—remote, independent, close-knit and apart from the world. The first hints I got of the gold mine were in stories of Caleb helping to outfit some of his friends who wanted to come to the valley but couldn't afford the move. I was told that they wondered a little at this, but all decided he must have had an extra good trapping season.

Then, when they got in Price, Caleb was always showing up with help of one kind or another, but still no one commented too much about it.

The valley was Caleb's idea, and he was in a way responsible for the success of its settlement.

So matters went along for a few years—until Caleb's dream collapsed. First, the Denver & Rio Grande railroad came through the town and they were no longer isolated. Then other settlers came, people Caleb had not invited. Then wagons began hauling freight to the Uintah Indian reservation and to the fort there. And finally coal was discovered. They were no longer a pioneer community.

He Turned Bitter

In the beginning, Caleb had been behind almost everything. He brought the meat for the first Thanksgiving. He helped plan and cut logs for the first meeting house. He was on the first school board. When the first Price ward of the LDS church was formed, he was chosen second counselor. There were those who said he was very disappointed that he was not made bishop, but he served for three years. In 1885 he quarreled with the bishop and resigned. He still had enough influence so the first counselor left with him, but his power was declining. Gradually he turned bitter and withdrew to his farm and his hunting, away from the town which had grown beyond him.

No one knows just when the first rumors began to go around that when Caleb Rhodes went hunting it was not only for meat, but for gold. It was back in the days when he was still the most powerful man in town. But no one dared ask him about it. As I've said before, he resented such questions. Several times one or another of the men tried to follow him, and once a group of them managed to keep him in sight for over 10 miles—but they always lost him. Whether he suspected that they were following or not, no one ever knew. He never said anything about it. But at any rate, he was too good a woodsman for them.

He Finally Confided

All they found out was that he headed toward Wyoming or in the direction of the Uinta basin. And when he came back he brought meat for his friends. But he didn't say where he had been. Later, when the trains were running, he would come back from one of these trips and take the train into Salt Lake. When he came back he'd have new clothes or furniture or a new kind of seed or a doctor or a newspaper man "talked" into coming to town.

"Rumors began that when Caleb Rhodes went hunting it was not only for meat, but for gold."

At any rate, Caleb finally confided in three of his closest friends, the two Horsley brothers and one of his Powell brothers-in-law. He told them that he had a mine over in the Uinta mountains, and he showed them dust and nuggets from the mine. He had sworn them all to secrecy first. When they asked why he didn't work the mine openly, he shrugged and said it was a long story.

The town grew up—and settled down. Most of the people built sturdy frame houses on the townsite. Their farms were producing well. They ran a canal through the town and planted trees. But Caleb Rhodes still rented out most of his acreage. He did a little gardening, raised a few honeybees, but spent most of his time in the hills.

His first wife and family had long since left him and he had married again. But still he could not settle down.

He Broke Down

Finally, one night in 1893, he broke down and talked to the two Horsley brothers about the mine. He talked until dawn. He was bitter by this time. He said that there was enough gold in the mine to pay the national debt (as of then) if he could only get permission to develop it, but that he could not. He said it was really ironic—that he had settled this remote valley and then coal had been discovered and the town had begun to turn from farming to industrial life, and all his dreams for it were gone. But because he had long ago signed an agreement, he could not develop his gold mine.

Caleb said that his father had discovered the mine back before 1868, and that it was in the reservation country of the Uinta mountains. There were two reasons why it had never really been worked.

First, it was in ground sacred to the Indians. They did not mind the two Rhodes men coming on it, as they had been taken into the tribe as full brothers. But they did not want unfriendly white men on their holy soil.

Signed an Agreement

Even so, the Rhodes men could have taken in a big enough force and managed to open the country and operate the mine. But at that time Brigham Young was opposed to mining. He did not want the type of people mine booms bring into a county brought into Deseret, at least not in its formative years. And Tom and Cecil Rhodes were faithful.

So they signed an agreement with Brigham which would allow them to work the mine themselves as long as they kept its location secret and only brought out a little gold at a time. They also agreed to give Brigham and the church a certain amount of gold. Caleb said that at this time he and his father were perfectly content with the agreement. After all, the Indians were their friends. And while it was hard work to get in and out of the reservation time after time to get the gold without being caught, they both loved the outdoor life anyway and would probably have spent much of their time in the basin country, mine or no mine. Also, they knew the mine was their own. It was only after Price was industrialized by the railroad and the coal mines that Caleb turned bitter. And even then he would not go back on his promise.

Secret Died With Him

He said that at one time they had tried to let one more man in on the secret, his half-brother, Enoch Rhodes. But Enoch was killed in the Uinta basin in 1883 by Indians close to where the mine was supposed to be.

When Caleb disappeared on his hunting trips he had, as so many suspected, gone again and again to the mine, each time taking out only such gold as he needed immediately.

"The Indians did not want unfriendly white men on their holy soil."

Those two nights, one in the early days and one when the town was built up, were the only times Caleb Rhodes was ever known to talk about the mine. But even then he did not even hint at its location, only to admit that it was in the Indian reservation country of the Uinta basin.

The secret died with him. Neither of his wives ever found out where the mine was. Indeed, they said afterwards that he never openly admitted to them that he had a mine, although they often wondered where all his money came from. He had three sons by his first wife. But two of them died in early childhood and the other one did not take to his father's way of life even before his mother and father separated. Even if Caleb had told him where the mine was, he undoubtedly would have been killed by the Indians if he had ventured near it.

So Caleb Rhodes died a bitter, lonely old man, none the happier for owning a gold mine which he could not work and which could not buy him happiness. He had seen his world, the world of the trappers and the mountain men and the pioneer villages, pass and a newer and, he thought, uglier world come in its place. The secret of his mine died with him, and I often wonder if he did not at least get a last ironic pleasure out of taking it with him.

Shootout Near Scout Lake

Author's Note

This story and the main story in the "Sky Pilot and the Gold" chapter are basically the same. I have included both to offer more information. They were told from different points of view, written at different times, and contain similar and diverse details. I hope you enjoy both.

Deseret News

Gold

November 26, 1989

Lee Davidson

Rumors of Riches: The search for the Lost Rhoades Mines in the Uinta Mountains has led many on a wild goose chase.

Old-timers say that somewhere in the Uinta Mountains are seven mines lined with rich, unbelievably pure gold that supplied the Aztecs with their treasures and spawned rumors about the seven golden cities of Cibola sought by early Spanish explorers.

Legend says that early Utah Indian chiefs who converted to Mormonism allowed Brigham Young to appoint one messenger—Thomas Rhoades—to be shown the mines and take gold for such church purposes as minting early Mormon coins and decorating LDS temples.

My grandpa, Amasa Alonzo Davidson, was one of hundreds who caught gold fever while hearing stories of the fabulous "Lost Rhoades Mines" and how Thomas Rhoades and his son, Caleb, rode out of the mountains with saddlebags full of pure gold ore.

Against his better judgment, Grandpa was talked into joining a group that searched for the gold in 1920. The trouble was that many in the party were outlaws, remnants of Butch Cassidy's wild bunch.

The stories of those who chased the gold are stores of misery. Some gold seekers were killed by Indians, some froze to death, some killed each other. And the few who reportedly saw the gold were prevented from claiming it because of untimely deaths or government red tape.

My grandfather's story starts with the story of Thomas Rhoades. Perhaps part legend, it's as real as I can reconstruct from relatives and books. It starts when Ute War Chief Wakara, or Walker as whites called him, was baptized into the LDS Church. In July 1852, Wakara agreed to let Brigham Young choose one white man to travel to a sacred mine and bring back gold if he swore not to reveal the location.

Young chose Thomas Rhoades, a stalwart Mormon who spoke fluent Ute.

In 1855, when Rhoades became ill, his son, Caleb, took his place.

Both Rhoadeses claimed they kept their part of the deal, but they also worked some not-so-sacred-but-also-rich mines that the Spanish had developed for themselves.

Thomas Rhoades found a map to the non-sacred Spanish mines when Brigham Young sent a group under his command to investigate an Indian massacre of Mexicans who had been mining gold near Nephi.

Neighbors began to suspect the Rhoadeses had their own mine and would try to follow them. When Thomas left his home in Kamas or Caleb left his in Price, curious neighbors would tag along, only to be outsmarted by the Rhoadeses.

Some claim they got close. Caleb once left a group in the Uintas for about five minutes and came back with a saddlebag full of gold.

One of the most determined seekers of Rhoades gold was Edward Hartzell—one of the men who ended up on my grandpa's expedition.

Once Hartzell thought he was finally hot on Caleb's trail without Caleb knowing it. But Hartzell had to dismount his horse and look for signs of the trail ahead in the moonlight. When he came back to the horse, he found someone had stolen his pistol. Caleb gave it back to him weeks later, saying he found it in the canyon.

After Caleb died, Hartzell married Caleb's widow. Some say he did that mainly to get information from her about the mines, but she didn't know where they were either.

Hartzell joined up with the same band of men as my grandpa, who tried to find the Rhoades gold by traveling into the Uintas from the Wyoming side.

In 1920, my grandpa was a 30-year-old rancher and schoolteacher living with his wife and five children near Fort Bridger, Wyo. One of the few in the area who knew how to assay gold, he was talked into joining the group.

One day the group rode up to Grandpa's ranch with an extra horse packed for him. When Grandpa saw some of the hard-looking characters, he refused to go.

All the men left except one of Grandpa's friends, Harold Mosslander, and the leader of the expedition, a self-proclaimed clairvoyant named Landreth, who had just moved to the area from Pittsburgh.

Landreth told Grandpa that he need fear nothing about the trip, and that he could prove it. He pulled out a sealed deck of cards and had Grandpa and Grandma shuffle and cut it. Landreth said spades were bad luck, clubs meant trouble, hearts were love and diamonds were riches.

Grandma cut the deck and drew the king of hearts, which Landreth said represented Grandpa. Grandpa did the same, drawing the queen of hearts, which Landreth said represented Grandma. They then shuffled the deck and drew four cards—the ace, king, queen and jack of diamonds.

Grandma let Grandpa go on the condition he leave his rifle at home. He rode out the next day with assaying acids, a blow pipe to heat them and a camera.

Shortly afterward, Landreth, while blindfolded, drew out a rough map that he said the spirit of an Indian princess named Ravencamp was revealing to him. He described the lake where they were to camp that night and said that above it appeared to be giant castles. The men in the group became excited because Landreth had described a place they knew well, even down to rocks that looked like castles.

In subsequent days, Landreth pulled out a compass that he said pointed toward the gold instead of to the north. He said it stopped working if the men didn't believe, and Grandpa supposedly was the biggest non-believer—which caused friction.

Landreth also could look at the men in the group and tell them things about their past that no one else knew. He even told one man he had killed and cut up a child. That kept the men in line, except for Grandpa and his friend Mosslander—of whom Landreth never seemed to discern anything.

Eventually Landreth led the group to an old cabin, which Grandpa later said was on the shores of what is now called Scout Lake by the Boy Scouts' Camp Steiner. After supper all of the men except Grandpa, Mosslander and a Wyoming neighbor named Ernest Roberts went off by themselves for about an hour. Grandpa told the others he had a sense of danger.

The next morning, Landreth told the others at breakfast that the Indian princess had led them to the cabin for a purpose. He said the gold they would soon find should be given to him to start a church. That angered many.

Landreth then asked that he be blindfolded so the Indian princess could help him draw a map for the final short distance to the gold. But his pencil did not move. Then he called upon God to direct him, but nothing happened. In anger, Landreth threw off the blindfold and announced he would find the gold himself.

One of the outlaws then took command. He placed another "old outlaw from Price" in charge of Grandpa, Mosslander and Roberts and told him not to let them get away from the cabin. He paired others off to go in different directions to look for gold, and he took Landreth with him.

During the day, the guard told Grandpa that Landreth and the outlaw leader the night before had told the others that Grandpa, Mosslander and Roberts should be killed as soon as the gold was found. They had even run their horses off. So he told Grandpa that no matter how rich any found gold was, he should tell them it was fool's gold. The outlaw also slipped him a gun.

At the Scout Lake camp, all the men—except two—returned before sunset but had found nothing. Just as it was getting dark the last two returned. The outlaw leader asked them what they had found, and one man said nothing.

Landreth yelled that he was lying and stepped toward him to search him. The man reached for his gun, but the outlaw leader shot him in the back. They brought the wounded man to the fire where Landreth searched his pockets, which were full of rich, gold-bearing rock—some of it almost pure.

The outlaw leader asked him where the mine was. The man said, "We found the mine over in—aw, go to [heck]." That outraged the outlaw leader, who shot him dead. He then ordered Grandpa to test the rocks with his assaying acids. He did and said they were worthless fool's gold.

"We found plenty of sulfur, some copper, some fool's gold—but no real gold."

The outlaw leader looked at the gold, and then looked at Grandpa. Slowly he drew his gun. But before he could fire, Mosslander shot him dead.

Everyone then started shooting. Grandpa and Mosslander ran out together, firing behind them as they went, and headed north to Wyoming. They said they ran so hard downhill that they even knocked over some trees. My Grandpa later found he had stuck one of the pieces of ore in his pocket when he started running. He donated it later to the University of Wyoming in the 1930s, but officials there said records are not good enough to verify that.

Grandpa and Mosslander found one of the horses that had been run off and headed home. Several men in the group had been killed, but many survived—including Landreth. Roberts returned home days later, wounded in the groin. The wound eventually killed him.

My father says the outlaws showed up at Grandpa's ranch and started digging around. Grandpa ran them off with the help of his brothers. Later, Grandma and her oldest son were shot at while working in a garden. She talked Grandpa into accepting a teaching job offered faraway in northern Wyoming.

Grandpa talked little about the trip to family members. He kept the gun he had been given in a box in hopes of returning it to the outlaw who had helped him. He never went back to the Uintas until 1951, and then only for one day. He had told his attorney about the expedition, and they decided to take a drive up and look around.

His journal entry says they found the old cabin and an old sulfur mine that was a landmark for him. We found out much later that he made a map of where he thought the two men who found the gold in the group were sent to look—a square area north of Scout Lake and just south of Gold Hill.

My father found the map while going through some of my grandpa's old papers. Grandpa in very light red pencil had traced in the route his expedition had taken. It is barely noticeable among the other red and black lines on the map.

My father, I and a Deseret News Photographer recently went to the area to see if we could see any signs there of the gold or the cabin mentioned by Grandpa. Where my father thought the old cabin stood, we found Camp Steiner's amphitheater for Scouts. It is made of logs laid on the ground. We noticed many of the logs are notched, meaning they were once part of an old cabin.

The site matches stories from my grandfather. He had taken pictures across the lake looking at Bald Mountain and Hayden Peak, and the view from the amphitheater matches those old photos described by my father. The site would also have allowed my grandpa to have run away during the shooting toward Wyoming in an almost all-downhill path.

We walked around and found the sulfur mine to the north that my grandpa mentioned as a landmark. We found plenty of sulfur, some copper, some fool's gold—but no real gold.

If any reader wants to follow my grandfather's map, feel free—but beware. Books detail many failed expeditions, misery and death and not one case of quick and bounteous wealth. Old-timers say for every ounce of gold in the mines, gallons of blood have been spilt.

But if you do find the mine, please let me see it some time. You keep the gold, I'm interested in another type of treasure—the treasure of seeing, feeling and even smelling such a rich mine, and knowing for myself it is real.

The Mysterious Dry Fork Cavern

The Vernal Express

The Mine of Lost Souls

Dec. 19, 1946

William Schaefermeyer Sr.

This story is of a strange and mysterious gold mine and cavern which is supposed to be in the vicinity of Dry Fork. The story is founded on truth and not on fiction or Indian legend, and its parts were related to me by men I can vouch for.

Fred Reynolds Sr. was at my home one day and in the course of our conversation I told him of some of my prospecting adventures in the Uinta Mountains. Mr. Reynolds seemed interested and said, "Let me tell you something." This is what he related.

One day I was over at Roosevelt and had occasion to exchange greetings with a group of men on the sidewalk. One of the men said to me: "Fred, do you know anything about the Dry Fork country?" I explained that I did—that I had lived there for a good many years, had herded sheep in the country and had hiked the area thoroughly. Another of the men then asked me if I knew of a good gold mine in the vicinity and I replied that I did not. They then asked me if there was a natural bridge near Dry Fork, and I replied that there was and that I had my picture taken on a horse on top of the bridge. I was then asked if there was a cave near the natural bridge and then after I had told them yes, I asked how they knew about the area.

One of the men then showed me a map and indicated that all the features mentioned were shown. The map had marks to indicate a gold mine, but it was old and many of the original markings were defaced through age and handling.

I gave little more thought to the incident until sometime later when I was herding sheep on Dry Fork, and a small boy came to me and said that there was a big hole under a cliff near by and that he had nearly fallen in while picking wild cherries. I investigated and found the hole, which opened into a cave, and was reminded of the map, as the cave was not far from the natural bridge.

Of course Mr. Reynolds's story interested me greatly, and although he explained that he found no gold, I was determined to investigate the story further. Mr. Reynolds told me to look for a Mr...... in Roosevelt if I cared to get more details.

One of my sons and I went to Roosevelt in search of Mr. H. . .and when we found him he said that he knew nothing about it, except that a Mr. Mitchell, who worked at the gilsonite mines at Fort Duchesne, had a map and told a fascinating story. We found Mr. Mitchell at his home and questioned him. "Yes," he said, "I brought the story up from New Mexico." And he related the following.

Mr. Mitchell's story

Some years ago I and my companion were traveling in New Mexico and as we were traveling along I suggested that in as much as it was getting late we should find a place to stay all night. We saw a Mexican hogan nearby and stopped and asked for a night's lodging—offering to pay. We were admitted.

In the evening we became the center of a small group of Mexican people and explained to them that we were on our way to our homes near the Uinta mountains. A very old Mexican woman sitting among the others flared [*sic*] noticeably when we mentioned the Uinta mountains and raised her hands to exclaim: "Oh, the Uintas, the Uintas," and she turned to us and said: "I tell you something."

"The map had marks to indicate a gold mine, but many of the original markings were defaced through age and handling."

Very long ago when I was a young girl, I, with some of my people, made a long journey to the far-off Uintas. We traveled over the Escalante trail. Near the end of our journey we came to a large stream (perhaps Ashley Creek) and we followed the stream into the mountains. We lived in a cave and each day traveled to a tunnel where we mined gold. (At this point she motioned with her hands to indicate a slope or pitch downward) and described with gestures the Sawtooth Mountains and a number of great pine trees.

Once we saw Indians watching us and we were afraid of them, so we moved from the cave to the tunnel. One day we took our gold and traveled far away to a little town called Spanish Fork, where there is a lake far off, to sell our gold. American men attempted to follow us back to the mine but we hid from them and made our way back to the mine.

One day my husband and I went out hunting and when we returned all our people were dead. The Indians had killed them. We were terribly afraid, so we put all our dead people in the tunnel, did some masonry work to wall up the opening and after covering the place with brush hurried away with all the gold we could carry. We traveled at night and slept in the daytime until we arrived in Mexico.

Mr. Mitchell's Story Continues

The Mexican woman then gave me a map showing the location of the gold mine, with trails, a cave, a natural bridge and a grove of big pines.

Upon my return to the Uinta Basin, I, of course, began to search for the gold mine. The map indicated that Dry Fork was the vicinity of the mine. There seemed no mistake about it, but my search has been fruitless.

I asked Mr. Mitchell if I might see the map, and he said that he had recently let it out of his hands, believing that he himself, was finished with the search for the mine. He said that it had passed through many hands and that it was now worn and probably illegible.

I then suggested to Mr. Mitchell that there were probably many men now looking for the mine. He said that such was true, but that no one had found it, as far as he knew.

On another occasion I visited Mr. Mitchell and urged him to tell me again his story. He did so, and it was the identical story he related before.

After my own investigation as a prospector—for I have examined the Dry Fork country personally—I believe that the gold mine is there. I have been over the trails which are nearly faded away, and have studied the geology of the area. The big stream is there, the cave and the natural bridge are there, the pine trees and Sawtooth mountains are there. But, alas, the gold mine! Where is it?

Some time in the future, when the sands of time have run much lower, the sequel to this story will be told, and the spirits of those dead persons who guard the gold mine will give up their secrets. Then the final chapter of the story of the Mine of Lost Souls will be told.

The Rhoades Mine

Author's Note

The author of this article is unknown to me. A typewritten copy came into my possession from my uncle Cleo Bascom. It had no name on it or any labeling to reveal its origin. The original document consistently referred to Tom Rhoades as Tom Rhodes. This is a misspelling as the text obviously refers to Tom Rhoades. I have changed the spelling in the passages below.

Unknown Source

Did Tom Rhoades Have a Gold Mine?

All my life I have heard of the gold mine from which Tom Rhoades was supposed to have gotten the gold which Brigham Young coined into money in the early days, but up until the time, about 30 years ago, when I purchased a place on the upper Provo, he was more or less a mythical personage. Then he became a reality, for if ever you wanted to hear an argument, all you had to do was go into Hoyt's store at Kamas and mention Rhoades and his gold mine.

Generally there were old-timers gathered there and whichever side you took, you had backing as well as opposition and those old timers were sincere in their convictions. I have statements from possibly 20 persons who knew Rhoades and spoke freely to me.

The time these Rhoades facts were gathered was about 1930.

According to those old-timers, Thomas Rhoades was the first settler in that region, going there in 1859. The valley was known as Rhoades valley but was later changed to Kamas valley.

Sure He Had a Mine!

Now for a few of the reactions: This one is from Mrs. Gines during the month of June 1930. She died Oct. 4 of that same year. "Sure he had a mine! Sure, I believe Tom had a gold mine." He had three large families and they never wanted for anything as long as they lived in this valley. He used to put a saddle on his horse, a pack saddle on his large mule, take a few quilts, some food and his gun and leave the ranch about sundown. He would go up Beaver Creek and be gone sometimes a week or more, but always came back during the night.

"We were lead to believe he was getting gold for Brother Brigham (Brigham Young), but in those days we practiced what we preached, which was 'Mormon Creed'—mind our own business.

"Now some said his mine was on Shingle Creek, others said Wolf Creek, and others said in Rhoades Canyon. One man said it was over on Rock Creek, but I don't know where it was. I do know he did get plenty of gold rock, for I have seen some of it."

I wrote Thomas Draper at Talmage, near Vernal. He answered on April 24, 1930, and said in part: "I do know Tom Rhoades had a gold mine for I have seen the gold—said it was on Rock

Creek, but I don't know—have prospected all over the country and last summer found some mighty rich gold. Going back to try again this summer, but I must put my crops in first.

One day I called on Cal McCormick and asked him if he knew Mr. Rhoades. "You mean Tom? Sure I knew him." Reminiscences come easy to old-timers, so he went on: "Now I don't know whether Tom had a gold mine or whether the Indians gave the gold to him to give to Brigham, but I do know he had plenty of gold rock which was mighty heavy. It looked like; well, when we used to make bullets and would spill a little of the hot lead, it would splash all over and looked just like that rock. It was very soft and had sort of wires of gold running all through it. Wire gold they called it, but it looked more like beeswax to me with the wax being gold and some of the honey was yellow.

"I used to help Tom load his mule when he would start for Salt Lake and it would make both of us grunt like [heck] to lift a small sack of that ore up and put it in the pack bag.

"No, really, I don't know where he got it, but he would be gone a week or 10 days, and say, he knew how to cover his tracks for no one could follow him beyond Shingle Creek."

There was a man who told me his name was Rouple Hyrum, I believe, he said, "Is your name Sharp?" It was. Then he said: "I heard you been hunting for the Rhoades mine. Correct?"

"No. I have been seeking information regarding the supposed mine Mr. Rhoades is believed to have had. Why?"

"Well brother, you're barking up the wrong tree. Tom never had a gold mine. Brigham called him on a mission to California and when he came back he brought a whole wagon load of gold with him, but he sort of snitched on the old boy, for he only turned in part and kept the rest for his own use. That was bad for him for when he left here he went into Southern Utah and things were bad. Grasshoppers ate up all of his crops, so he got in a wagon and drove up to see Brigham and told him about the gold he had given him and asked Brigham to give him a load of flour and potatoes from the tithing fund.

"If ever you wanted to hear an argument, all you had to do was mention Rhoades and his gold mine."

"Now if Tom had known where there was a gold mine, do you think he would have gone back to his hungry kiddies without some food? Not by a darned sight. He would have loaded his wagon with gold and traded it for food—bet your life he would."

Here is another one, but I won't tell the name of the man who told it to me. Certain of his relatives living in Kamas might not like it if I did, so here it is.

"Tom never had a gold mine. Went to California and came back with a little Indian boy he called Sam—Sam Rhoades. They walked all the way and drove a little burro that carried their food and a bushel sack full of gold dust. Now I know this for I was there when Tom came back and took the pack off his burro, and saw that bushel of gold dust."

Here is the Ketch

"Now here is the ketch. If the figures given me from the assay office be correct, then that bushel of gold dust would have weighed 1,514.258 pounds. Some burro to carry that load and some man to lift that weight from a burro, but in those times they did really grow superhuman beings."

Richard Lambert at Kamas was a rather retiring man. He did finally say he did not know whether Rhoades had a gold mine or whether he was getting the gold from the Indians, but he thought the latter was true and that the gold was used for the church here in Utah.

Just one more. My mother came to Utah in 1861. She was at Woodland while I was gathering data on the Rhoades mine. I asked her if she had ever heard about Rhoades and his gold mine. She had but knew nothing about it, but did say this: "We lived in the eighth ward for many years. Shortly after we came some of the younger men used to disobey the church authorities and go away prospecting. Two came back late one fall with some almost pure gold rock. The bishop was inclined to disfellowship those boys for disobeying the orders, which were: they were to stay on their farms and raise crops, but it was hushed up.

And now for some reactions regarding the Madre de Oro story. These came from two different men, but are closely connected and cross each other's trails, so let me say that Arlin Davidson of Meadow, Utah, and Lloyd Henry or 1457 East 3010 South, Salt Lake City, told me these stories which I have combined into one so as not to have a repetition.

One of those men told me that when the bodies of those Mexicans were found down on Chicken Creek that Tom Rhoades was one of those men who made the discovery. That about all they found was the skeletons and some arrows which were buried. However, it is reported that one man picked up a small metal box or case, inside of which was a very old map written in Spanish and written on buckskin. This Rhoades is reported to have taken.

After Rhoades had the Spanish changed to English, he learned it was a map leading to a rich gold mine. He tried to find it by crossing the Provo River way up high, but could not find the mine.

Rhoades is reported to have had a son who was anxious to get enough ahead so he could marry the girl of his choice and in some way got a copy of that old map from his father and that he went in and started searching from the eastern side, taking the old Indian ford of the Green River, near Jensen, as his starting point. People saw him riding his horse and leading a pack animal, slowly riding to the west.

Apparently he found the gold mine, the Madre de Oro del Uintah, but when he stopped and picked up a piece of gold-bearing ore to examine it, an Indian shot him with an arrow. He slipped the gold in his pocket and had strength enough to mount his horse and ride away, but left the pack horse behind.

After he had ridden quite some distance, he began to grow faint from loss of blood so he dismounted and, taking an old envelope from his pocket and using a match for a pencil, dipped it in the blood and made an X and wrote, "I'm here." Then in a winding course he made a line to a place and made another X and wrote, "Indian—"

A few days later some stockmen found his horse wandering around and, seeing dried blood on the saddle, backtracked the animal. They found the badly decomposed body of a man. They searched his pockets in an effort to find out who he was but found nothing but the piece of gold he had found. However, they did find the map nearby. After looking at it for some time they decided whoever had made it had wanted to tell where the Indian had been when he was shot. Back to their camp they went, secured a shovel, buried the body along with that piece of an envelope, and returned to their camp.

A few days later they notified the officers, who identified the horse as the one young Rhoades had been riding. They went to the grave and dug it open, but found the writing on the paper had all been obliterated so they covered the remains over and set about to notify the Rhoades family.

In due time those officers turned over that piece of gold ore that had been found in young Rhoades' pocket to the girl he was intending to marry.

Now there was a sort of love affair showing up on the horizon. It appears that Mormon Selman had long sought the favor of Tom Rhoades in an attempt to get him to tell the location of his mine but had failed. Next he is reported to have made love to the young lady Rhoades Jr. was intending to marry and won her for a wife.

One of those men sort of scoffed at the idea that the Indians had given Selman that large piece of gold he had shown me and suggested to me it might have been the same piece of gold that Rhoades Jr. had found and which had belonged to Mrs. Selman.

The Sacred Mine

You have probably realized by now that yes, indeed, Utah has a rich heritage. I believe that many mines were worked by the Spanish in what is now Utah. A few of these mines had a very specific purpose and significance. The following story is about one such mine.

The bishop was a friend of Cris's who worked for the government as a surveyor. He enjoyed hiking, exploring, and the fresh mountain air. He had found symbols before, but not in the high concentration found in and around Paradise Park. These symbols were very old and were on some of the oldest pines he had found anywhere in the mysterious Uinta Mountains. It would have taken at least two men to reach around most of these large trees.

He first drove past Paradise Reservoir in 1939 and made it to this remote area deep in the Whiterocks River basin. The only thing he regretted was being away from his family for long stretches at a time while he surveyed and photographed for the government in these remote locations. However, the intrigue of the Spanish history and the meaning of their symbols were irresistible. As he explored, he took many pictures which would later be labeled as the property of the United States Forest Department.

The area was fascinating and there were new discoveries to be made every day. An old Spanish mine mark with the date 1780 was very impressive, but even better was finding a Spanish cannon. It was obvious the Spanish had been there and had spent much time in the area, but why? It didn't add up unless there had been a rich mine nearby. A cannon in such a remote, inaccessible area could only have been used for protection from the Nativ Americans. He looked for the mine every spare moment but couldn't find it. The longer he studied the evidence, the more convinced he became of the existence of a hidden mine somewhere nearby. He was further convinced that it must be a rich one for the Spanish to go to the trouble they had.

As he contemplated the depth of his bewilderment, he came up with a possible solution. He would ask the medicine man for help. Over the years they had become close and trusted each other. Because they were friends, the shaman agreed to show him the location of the sacred mine.

His anticipation ran high the day they hiked to the mine. They walked the same trail the bishop had walked so many times before. Then, nearing the end of the trail, the shaman directed him easterly. Down over the lip of the hill they went. They traversed several boulders as they started down the eastern slope of Paradise Peak.

There, not far from the top, the guide located the particular boulder he had described to the bishop as they had walked along the path. He directed the bishop where to go to find the door. The bishop located the door and opened it, exposing some stone steps.

This was the moment he had so long awaited. He would finally get to see what very few had ever seen. He paused only long enough to light his lantern. As he walked down the stone steps, he noticed he was entering a room. His heart was beating with excitement and he had to take a deep breath as he allowed his eyes to adjust to the lantern's light and take it all in. He noticed a small table to one side with a very old copy of the Book of Mormon resting upon it. Along with the old book there were discarded whiskey bottles on and under the table. As he gazed around the room he noticed a large black candle secured to the wall. There were two unlocked, oversized wooden doors which next drew his curiosity. As he pushed them away from him, his view was opened into a tunnel. The shaman had told him this tunnel connected this mine with the next sacred mine to the north. As he thought about it and walked a short distance into the tunnel, he calculated it would have to be nearly two miles long. What an undertaking that must have been! It was cold, dark, and very wet in this tunnel. Water dripping from the ceiling fell on his head and the water on the floor increased in depth as he walked deeper into the black hole. It wasn't long before he decided he had seen enough and turned around and started out.

Even though the tunneling was nearly unbelievable, what he found even more mind-boggling were the veins of gold. Everywhere he looked he saw them. Some were in sheets and of one hue of yellow while others were more crystallized and had a higher silver content, which gave them a different color and look. There were others which were more of a chocolate color. He had been warned not to chip any gold from the main vein on the wall in the room because it had been poisoned. He complied with the request. However, he couldn't resist taking something to prove he had been there, so he picked up a loose piece of gold from the floor. It had a beautiful, eye-catching butter yellow color.

Having seen what he had come for, he made his way back up the stairs and out to his waiting friend. They climbed back up the slope and sat down to talk. He asked his guide why this mine was called the sacred one. The shaman explained, "It is called the sacred mine because they believe it is where the Great Spirit dwells. You see, in the next life they believe everything is made of gold: the leaves, the trees, and even the grass," he continued, "each has a different look, color and texture. The leaves are like the sheets of gold, and so on."

It all made sense to the bishop now. This is a sacred place because it reminded the Indians of the next life. Now he understood.

Words were hard to find as he thanked his guide. They had become close over the years and they had seen many great and incredible things together. He would never forget their friendship or the day he visited the sacred mine of the Utes.

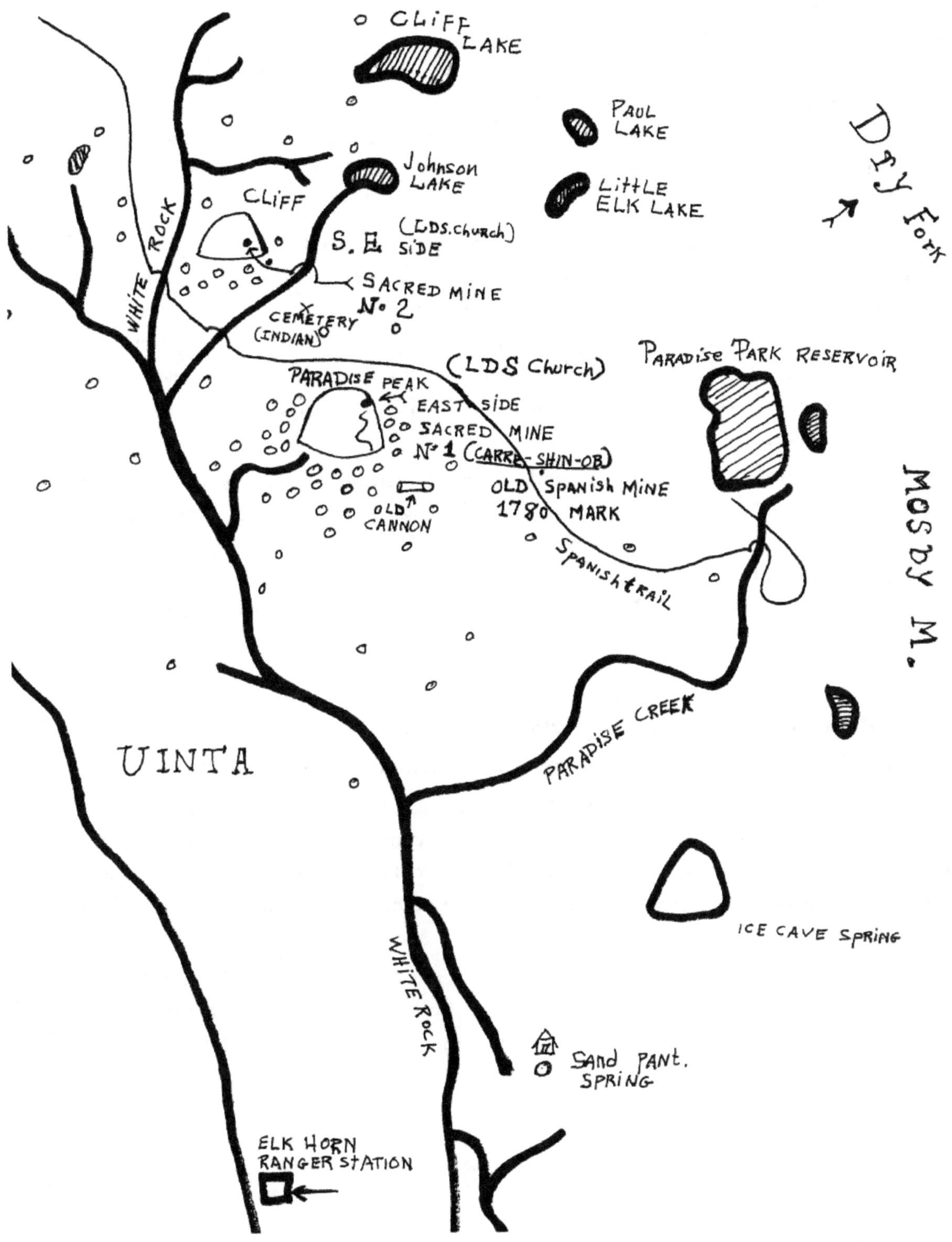

The origins of this map are unknown to me. Cris and I purchased it from the bishop. It's a map to the Sacred Mine or Carre-Shin-Ob that the bishop visited. He said he never would have found this mine if he hadn't been taken to it by the shaman, owing mainly to its concealment. As you examine this map you will notice two entrances or mines, marked No. 1 and No. 2.

A Note about Dowsing

As I explore and investigate various areas for clues of our past, I utilize tools which are within my budget and which appear to be effective. Dowsing is a method of seeking water, minerals, and other specific objects using a Y-shaped piece of wood or L-shaped metal rods. As far as I am aware, dowsing hasn't been scientifically proven; however, I have had some success with it and I am willing to share a few results and insights with you.

Some dowsers appear to have a talent for dowsing for minerals and water. My father, Raymon Bascom, is one who is well known for dowsing for water. His name is known from Utah Valley to parts of Colorado. I would estimate his accuracy to be better than 95 percent for finding water. He can tell you where underground water is on your property, which way the stream flows, how deep it is, and how many gallons per minute it flows. He also dowses occasionally for minerals. I have been with him twice when the rods have led us more than half a mile around hillsides to iron deposits. That alone is enough to convince me that they work.

Others also have this talent. A friend of Cris's father, Pedro Angulo, has the ability to map dowse for veins and minerals with surprising success. Cris and I sent him a map of the entire Whiterocks drainage area and asked him to dowse for the Sacred Mine. He did and then sent us the results, which I am willing to share with you. Out of the entire Whiterocks Canyon, he picked an area close to Paradise Peak as the location for the mine Cris and I were looking for. This correlates fairly closely with the information and map we purchased from the bishop.

Explanation of the Following Dowsed Map

These are Pedro's words as Cris translated his explanation from Spanish to English for me:

What we can analyze on this map is something big. I have interpreted two different areas. The first one, we call mine number one and is in the circle. We can't determine the size of this mine because what is inside is so big. This indicates a huge deposit of gold and silver. The mother lode is so big the minerals have passed from one side of the fault to the other. I also detect a lot of water inside this mine.

Mine number two appears to contain huge portions of metal, both gold and silver. These mines are connected by a tunnel. Also I found a lot of water in both mines. Both are very rich in gold and silver. I also marked, with an X, where a cemetery is because there are a lot of human bones lying around. It is worth it to investigate this place. The circled X is another mine of gold, only smaller.

Sincerely,

Pedro Angulo

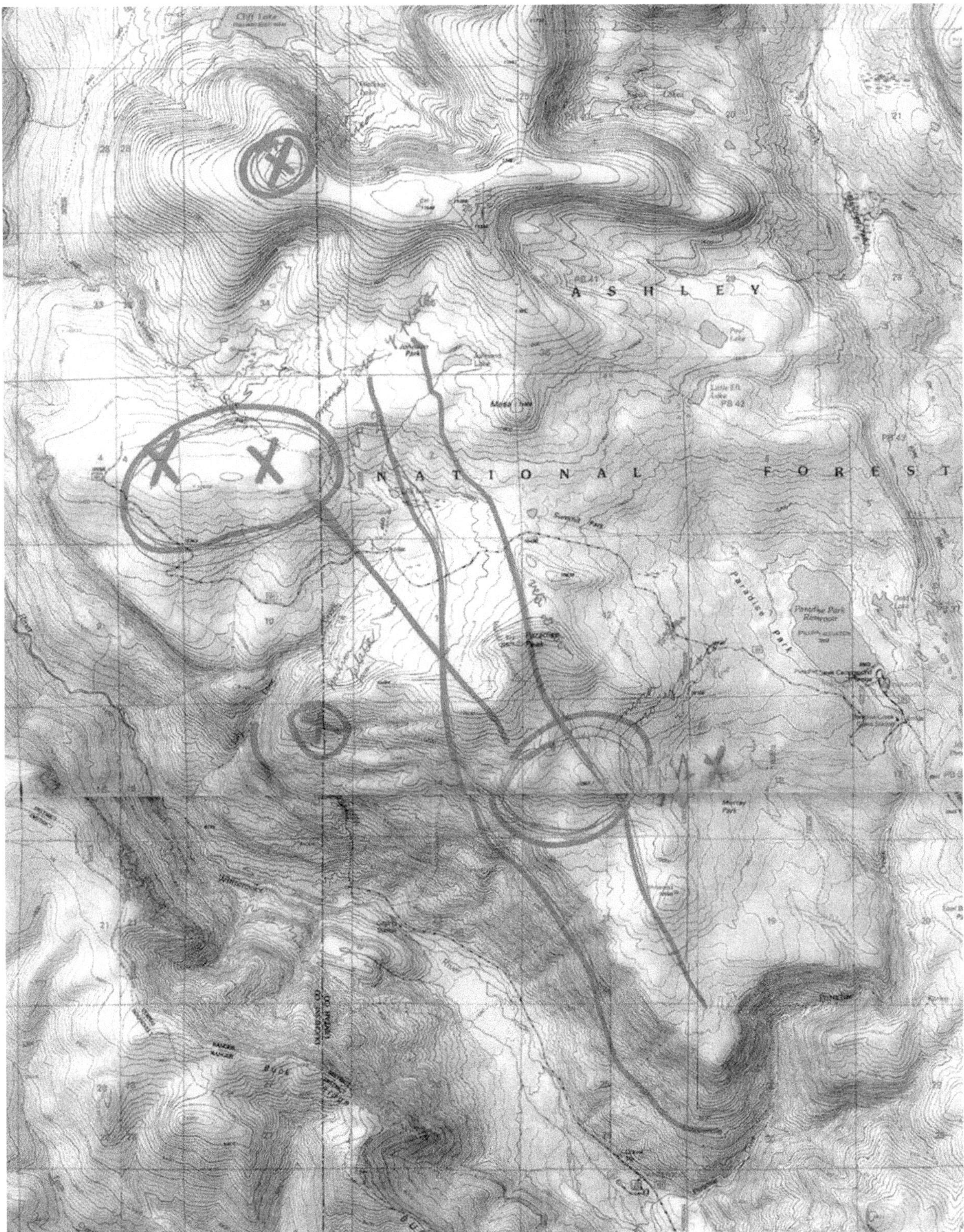

This is a map of the Paradise Peak area Pedro dowsed. The locations and ideas Pedro came up with are similar to the bishop's map. Map created with TOPO! © National Geographic Maps.

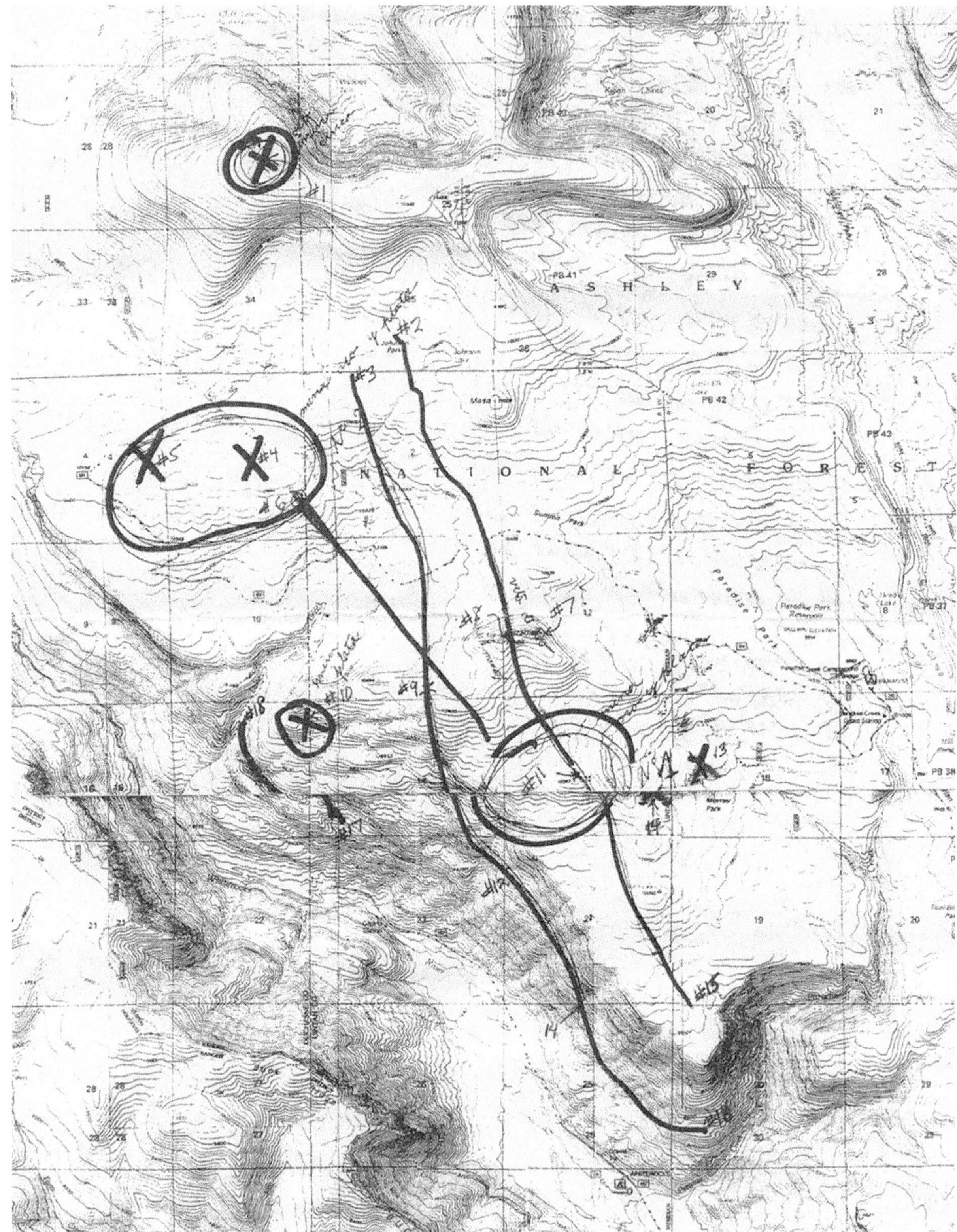

Here is the same map with approximate GPS coordinates to help locate these spots more easily while hiking. Map created with TOPO! © National Geographic Maps.

GPS Information

Author's Note

Many GPS coordinates are given throughout this book. Unless otherwise noted, they will be offered in the NAD 27 Alaska format. Coordinates are given to enable you to verify my findings and to give you a head start.

Paradise Peak Area

Estimated GPS Coordinates from Pedro's dowsing (only estimated!)
(NAD 27, in degrees, minutes and seconds)

1. 40° 42′ 47″ N. 109° 58′ 52″ W. "A Little" Gold
2. 40° 41′ 50″ N. 109° 57′ 59″ W. North End of East Vein
3. 40° 41′ 38″ N. 109° 58′ 16″ W. North End of West Vein
4. 40° 41′ 11″ N. 109° 58′ 59″ W. Sacred Mine No. 2
5. 40° 41′ 11″ N. 109° 59′ 40″ W. Another Possible Entrance
6. 40° 40′ 56″ N. 109° 58′ 37″ W. Underground Passage
7. 40° 40′ 13″ N. 109° 57′ 00″ W. Sacred Mine #1 according to the bishop
8. 40° 40′ 14″ N. 109° 57′ 13″ W. East Vein
9. 40° 39′ 57″ N. 109° 57′ 47″ W. West Vein
10. 40° 39′ 45″ N. 109° 58′ 38″ W. Silver Mine
11. 40° 39′ 28″ N. 109° 56′ 46″ W. Sacred Mine #1 according to Pedro's Dowsing
12. 40° 38′ 51″ N. 109° 57′ 16″ W. West Vein
13. 40° 39′ 31″ N. 109° 55′ 56″ W. Scattered Bones
14. 40° 38′ 09″ N. 109° 56′ 44″ W. West Vein
15. 40° 38′ 09″ N. 109° 56′ 01″ W. South End of East Vein
16. 40° 37′ 29″ N. 109° 55′ 56″ W. South End of West Vein
17. 40° 39′ 08″ N. 109° 58′ 25″ W. South End of Smaller Vein
18. 40° 39′ 47″ N. 109° 59′ 05″ W. North End of Smaller Vein

Things I Have Found

40° 40′ 13.8″ N. 109° 57′ 4.4″ W. At this location, I found a hole drilled into a rock, where it appears survey equipment was used. There are also some spikes driven into the dirt which, it appears, anchored the pole. The bishop may have set up his survey equipment here.

There is a distinct trail on Paradise Peak that looks similar to the one on the hand-drawn map on page 57. It hasn't been groomed in quite a while. However, it was well-built and is easily followed. I followed it to its end and that is literally all that happened. The mystery here is which way should I go after reaching this dead end?

I found this unique trail while looking for the Sacred Mine.

This is a picture of my sister Jerry Ann, her husband Ken Lance, and their boys Blake and Burkley, resting at the end of the trail.

The photo above shows what I call the Paradise North Mine, located at 40° 40′ 14.8″ N. 109° 57′ 0.4″ W.

I stumbled upon these mines while looking for the Sacred Mine. Neither of these appears to be the Sacred Mine.

I call the mine pictured here the Paradise South Mine, located at 40° 40′ 8.7″ N. 109° 56′ 56.4″ W.

Markers Near the Sacred Mine

The following symbols were catalogued and photographed in 1939 by the bishop. Upon his return in 1940, nearly all of the large trees with symbols carved into them had been cut and hauled to the mills for lumber. He said the trees were large enough that it would have taken two or more people to reach around them.

I apologize that I wasn't allowed to copy the pictures or show you the original photographs the bishop took. The next best thing I can do is show you what they looked like through drawings. Cristian Vera spent many hours studying the original pictures and reproducing the symbols as accurately as possible. We both hope you enjoy the study of these symbols. Maybe you will be one of the few who can decipher them.

One day, while closely examining one of the unique symbols in this area, the bishop noticed something wedged into a crack within the symbol. Since then, the tree had nearly overgrown the object. He pulled out his pocketknife and started digging to see what it was. He ended up with a Spanish eight reales coin. This lends weight to the hypothesis that these symbols are Spanish in origin.

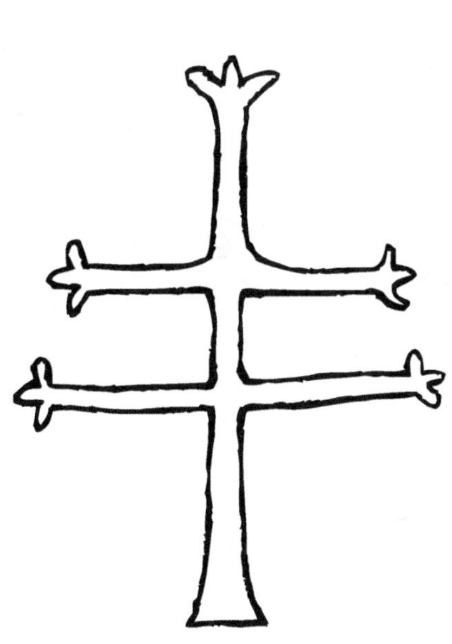

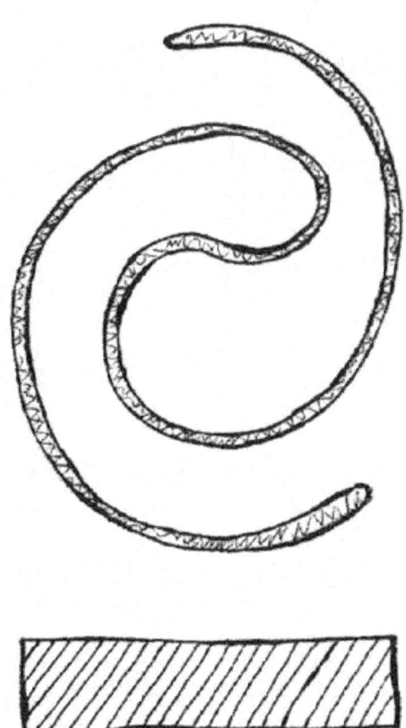

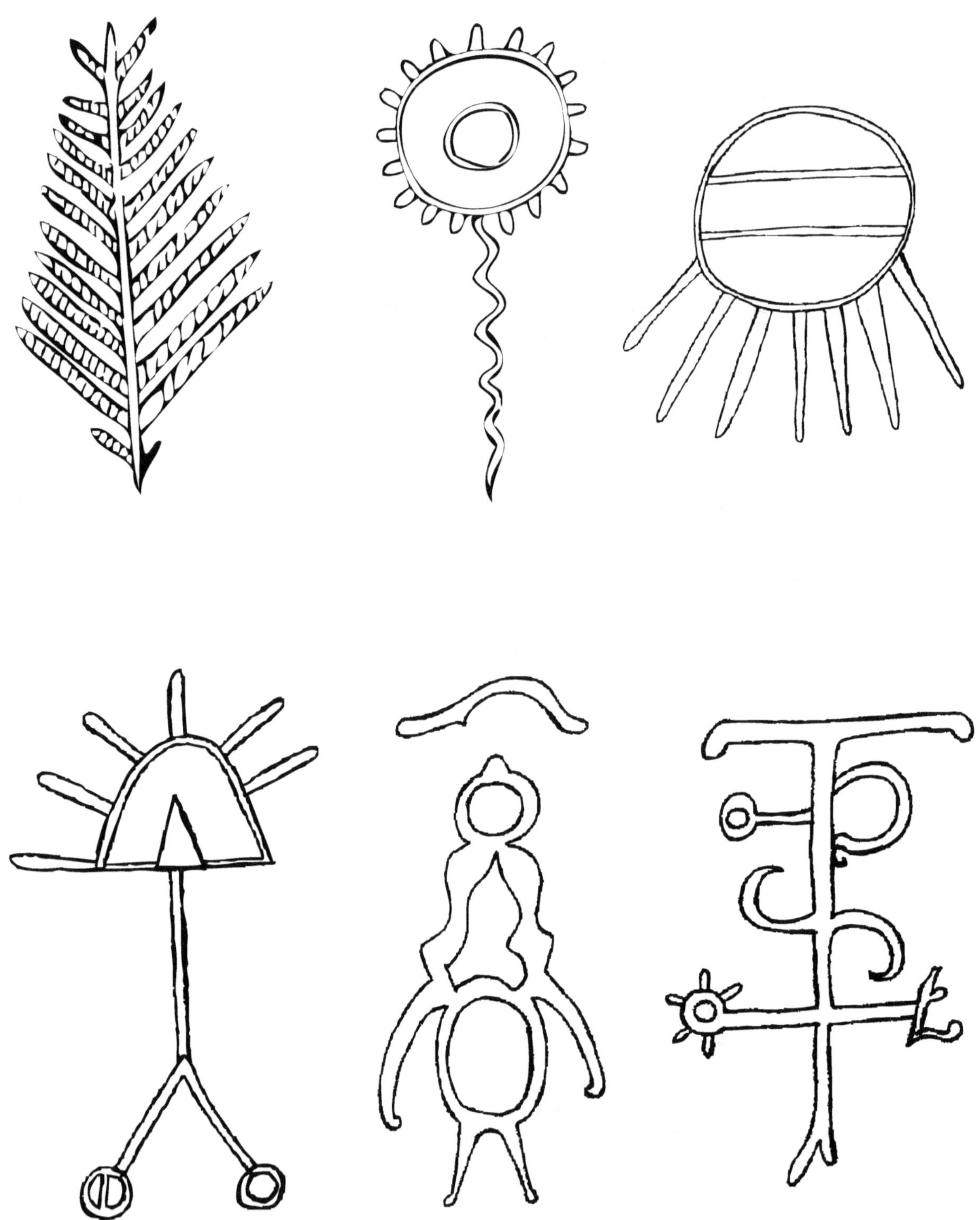

hole through
tree trunk

N

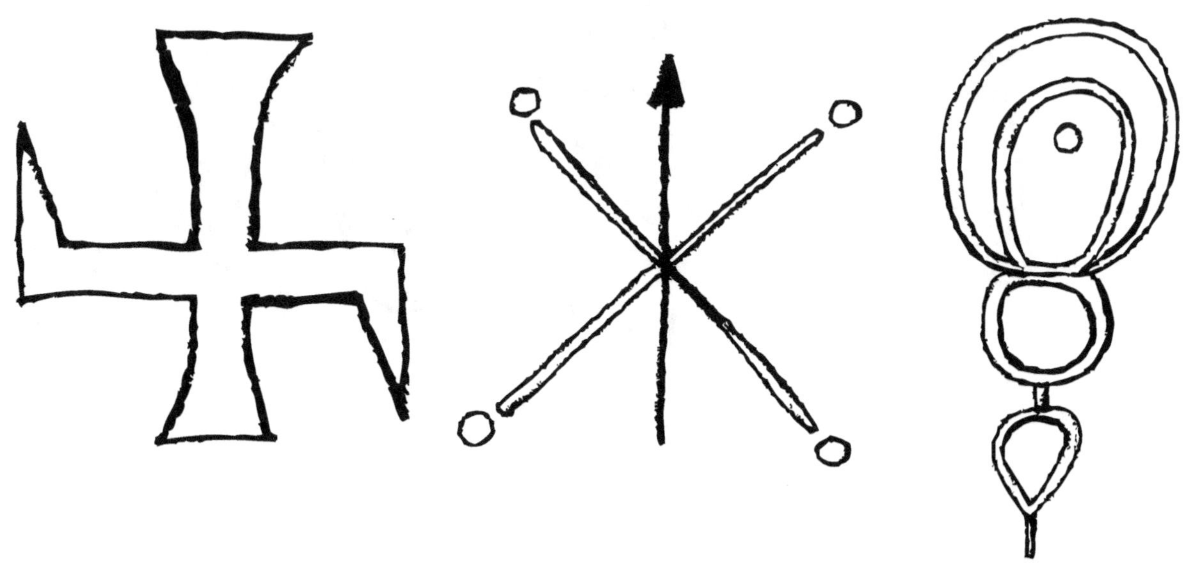

FER
VII

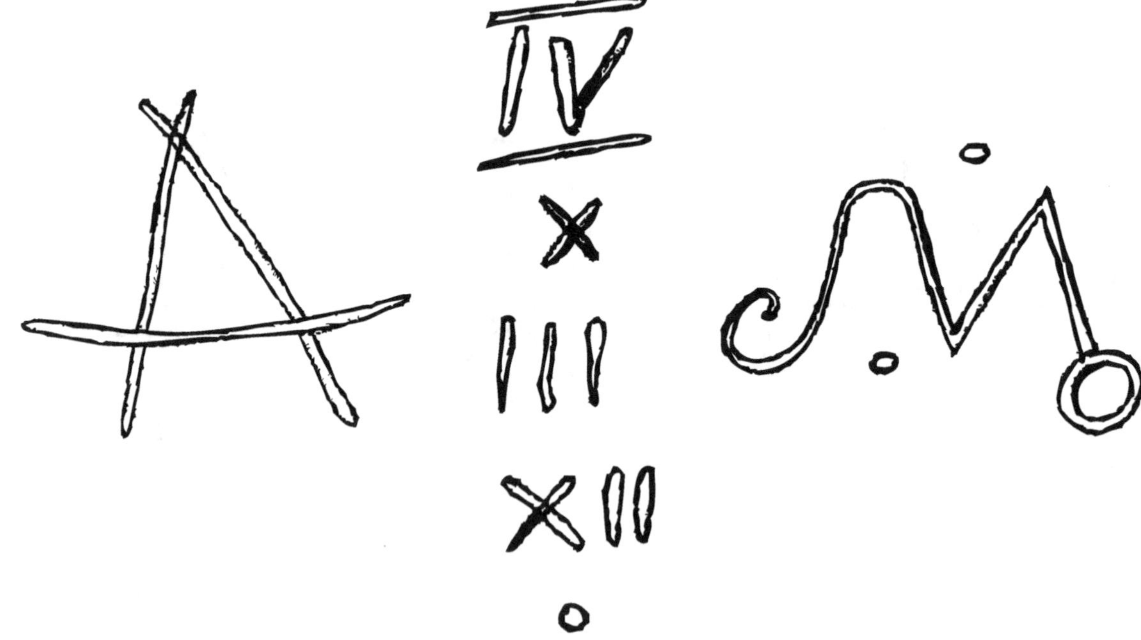

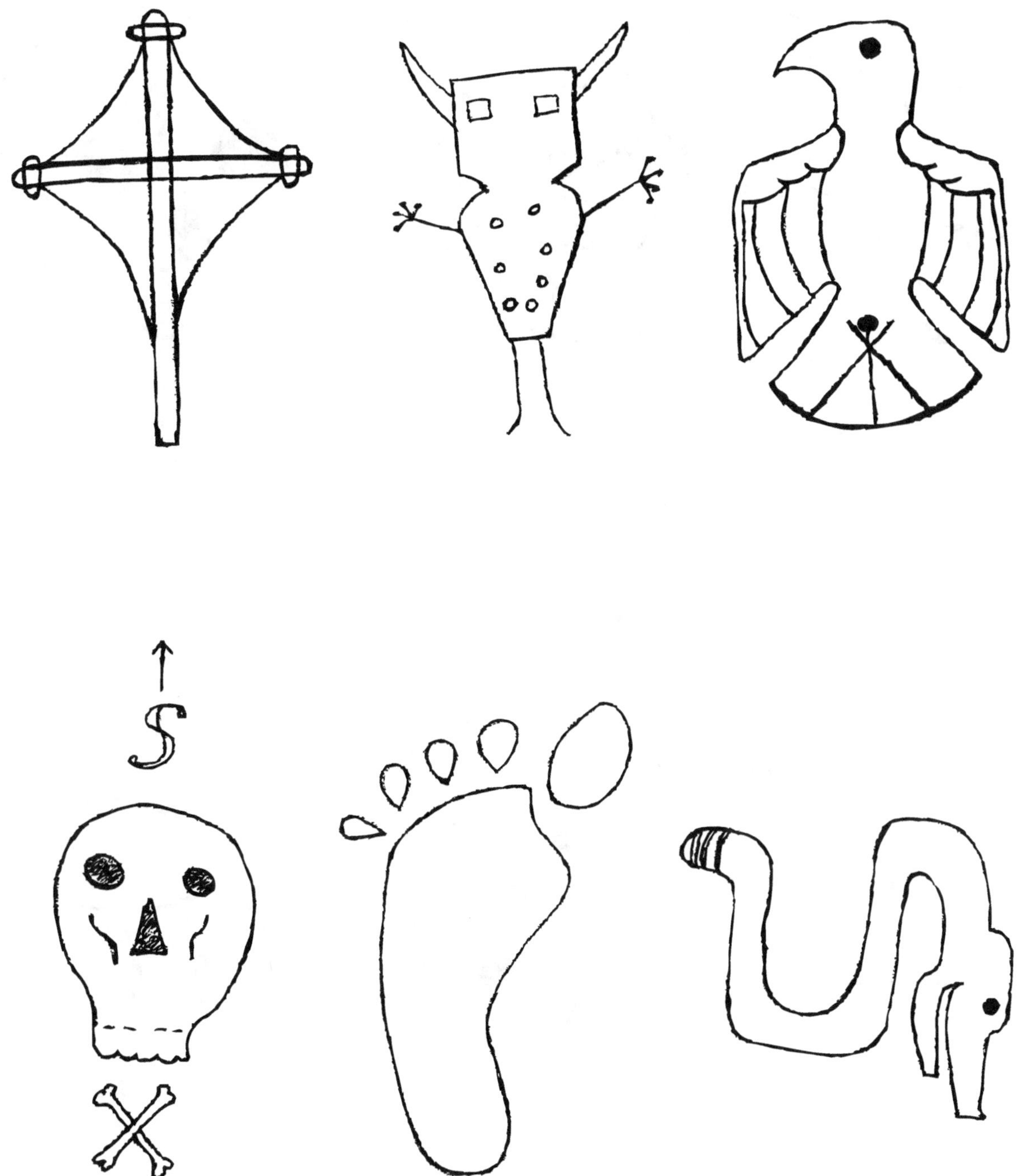

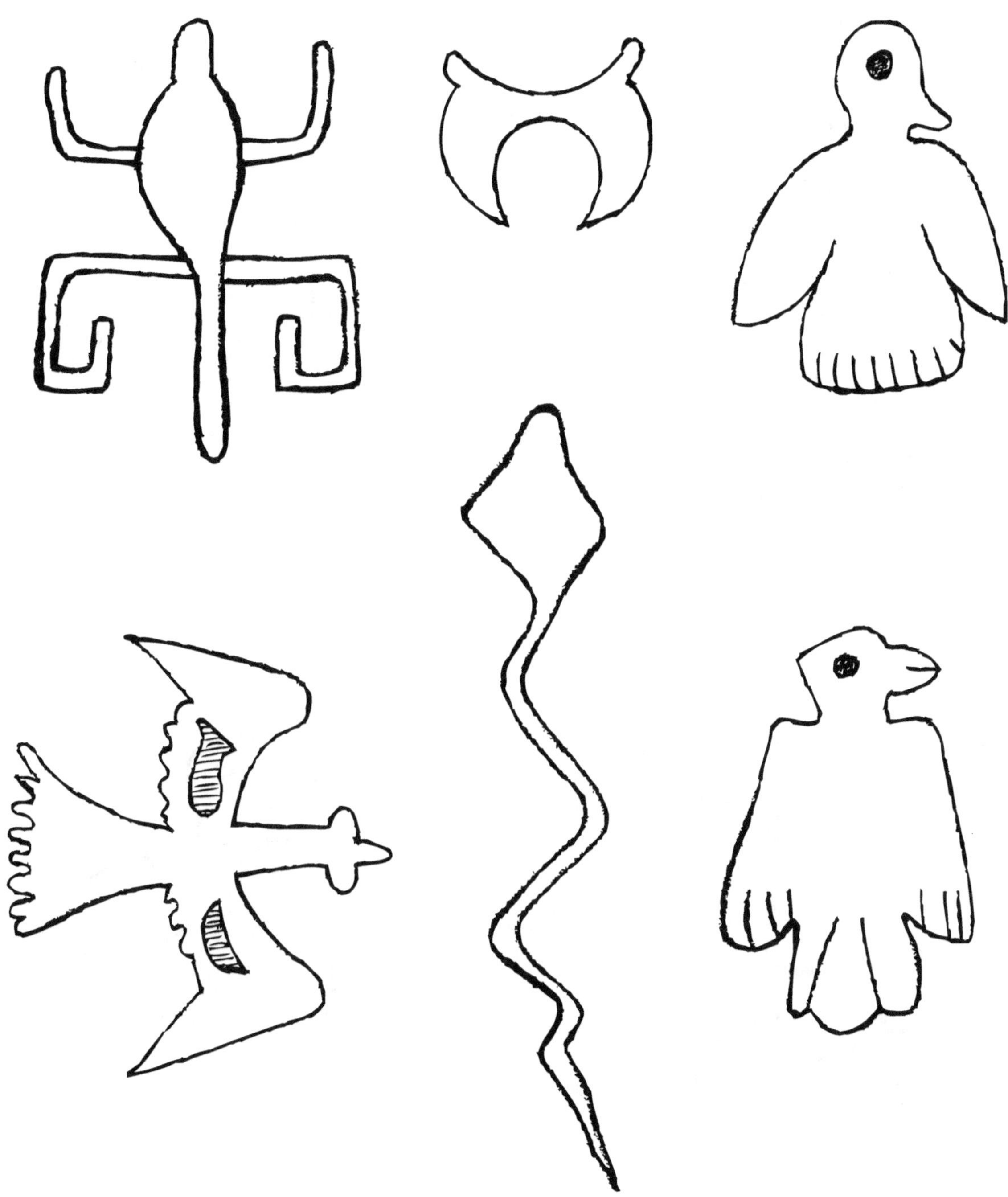

Translation from Latin:
Health or security of the people.

Translation from Spanish:
Bad Indian.

A Spanish surname.

Translation from Spanish: Cactus.

Translation from Spanish:
Spirit.

Translation from Spanish:
No Trespassing.

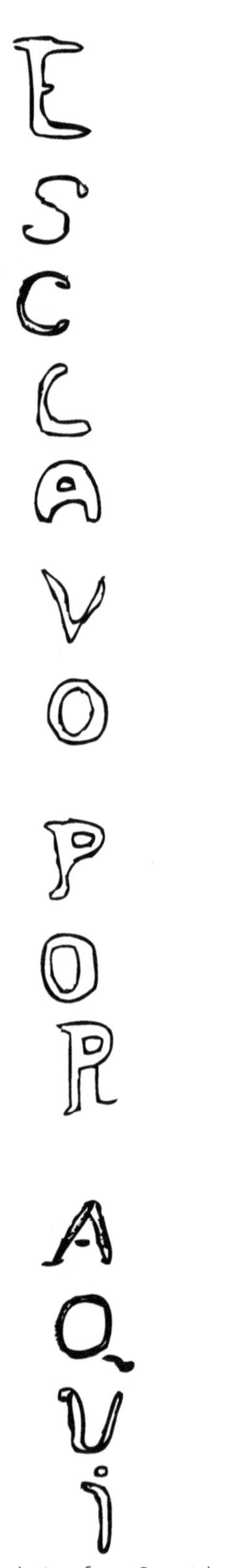

Translation from Spanish:
Slave this way.

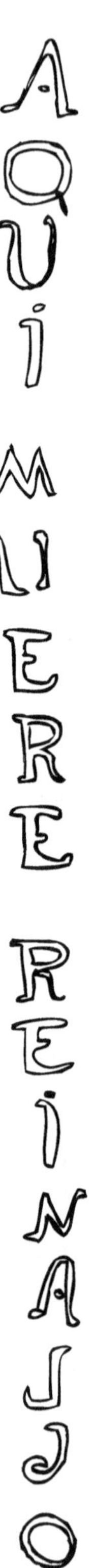

Translation from Spanish:
Reinaldo died here.

More Markers and Symbols

Author's Note

These are photographs of Spanish and Native American markers and symbols found throughout Utah. The highest concentration was found in the Uinta Mountains.

Many people, including myself, have driven past this roadside boulder in Brownie Canyon without knowing it had symbols on it. It's unknown whether the symbols are Spanish or Native American. We do know that by interpreting them as though they were of Spanish origin, my dad and I have found mineral deposits.

The sun symbol is at 40° 04.773′ N. 111° 24.155′ W., elevation 6072 feet.

This is a close-up of carvings found on a large pine tree. It's at 40° 38.912′ N. 109° 45.815′ W., elevation 8060 feet. It's nearly within a stone's throw of the road in Brownie Canyon.

This is Birdie Vernon and her grandson in front of the same tree.

"Don't follow the crosses; instead follow the bird symbols," the bishop said.

These symbols once greeted us as my dad, my sons, and I arrived at Horse Shoe Park. The photos are from a year when we arrived quite early in the season. These pictures were taken around 1985.

A few years later, my dad and I went back and followed these and other blazes to the north and east and ended up on the south slopes of Marsh Peak. We were tuckered from our long hike. As we neared the mountain, we found a well-used logging road that we didn't know existed. The next time we go into this area we will take the easier way!

The bishop later told us we should follow the eagle or bird symbols and not the crosses. He said the crosses usually lead to cemeteries. He also said the eagle symbols often lead to the richest mines.

My son Travis tried to keep up with us as we waded through the ankle-deep water from the snowmelt.

These symbols were photographed on September 3, 2005, at Lost Park, which is near Paradise Reservoir. We may know what the crosses mean, but what about this split cross symbol? Your guess is as good as mine.

Have these trees been altered to become markers?

What is your opinion?

It looks like this limb was tied in a knot when the tree was very young. Through experimentation, Dan Lowe discovered that the limb would have been smaller than three quarters of an inch in diameter to keep from breaking at the time the knot was made.

One day my dad told me about some symbols he had found in Whiterocks Canyon. Together we drove to the location to take a closer look. We were unable to locate one of the symbol trees beside the road, but the other one is pictured here.

If we interpret the symbols correctly, the bottom symbol represents the top of Buck Ridge and the ridges running east from it. The jog represents the jog Buck Ridge makes, and the dot would be a mine. We took a hike to see if our theory was correct.

We found many more symbols and other interesting things, like the source of most of the water in that small side canyon. It gushes from a hole about one foot in diameter and can be heard for a considerable distance. We didn't find the mine. Every time we wanted to go back and look again, it always rained on us. There is a waterfall in that canyon, but we didn't think to look behind it for the mine while we were there. We remembered later that this was a clue Caleb Rhoades used when describing one of his mines. We will return someday to look a little harder.

This is one of the symbol trees my dad found in the mid-1980s on the west side of the White Rocks Canyon road. To see a few of the symbols I found in this area, go to the map on page 115.

The rock monuments in this photograph can be seen for miles and could be one way the Spanish explorers and miners returned to their mines or navigated the mountains. It can be found in the area of Low Pass, which lies at the head of Current Canyon.

The symbols on this tree appear to be Spanish. The tree is dead and could fall over at any time. This marker is also in the Low Pass area.

The Cat Face Symbol

There is a very common symbol found in the Uinta Mountains which I have affectionately nicknamed the cat face. It has been too long to remember who first called it that, but the name stuck. As you examine the symbol in detail, it is easy to see the resemblance and the reason for the name.

While the cat faces themselves aren't difficult to find, their history is a little more elusive and difficult to pin down. Rock Harrison, in an email to me dated March 8, 2005, stated, "The cat face symbols . . . first started to appear here in the basin around 1977." He may be correct. While exploring in the early '80s, I found one on a tree on the Whiterocks side of Mosby Mountain. There was a "79" carved above the cat face. It appeared as though that could have been the date the symbol was carved into its bark.

There is a cat face that has fallen from a large quaking aspen tree in Rhoades Canyon. According to Verl Iorg, it once had the date 1911 carved next to it. This information may conflict with Rock's statement. I first saw the symbol and the tree in the summer of 2003. The large tree was dead and the bark with the cat face on it had fallen from it. I didn't think of taking the symbol at the time but returned a year later to do so, only to find it had deteriorated so much that it was almost unrecognizable and hardly worth taking. This picture shows what it looked like in 2003. It was near the base of the tree it fell from at 40° 30.162' N. 110° 53.778' W., elevation 9388 feet.

Usually cat face symbols are found carved into trees; however this one, which we found in Dry Fork Canyon, was carved in to sandstone.

It's my belief that the cat faces are markers to lead one back to the mines once worked by the Spanish and later the Mexicans. At many cat face locations I have looked at the symbol, walked past it, and found a mine—usually within one hundred yards of the symbol. A theory I have considered is that miners of Spanish descent may have checked on their mines, re-marked them with the cat faces, and are awaiting the time when they will be allowed to work them again.

Crosses

As I have explored the Uinta Mountains, I have discovered several crosses; all appear to be very old. Some of them were on rocks or ledges, while others were carved deep into pine or quaking aspen trees. There are several pictured in this book.

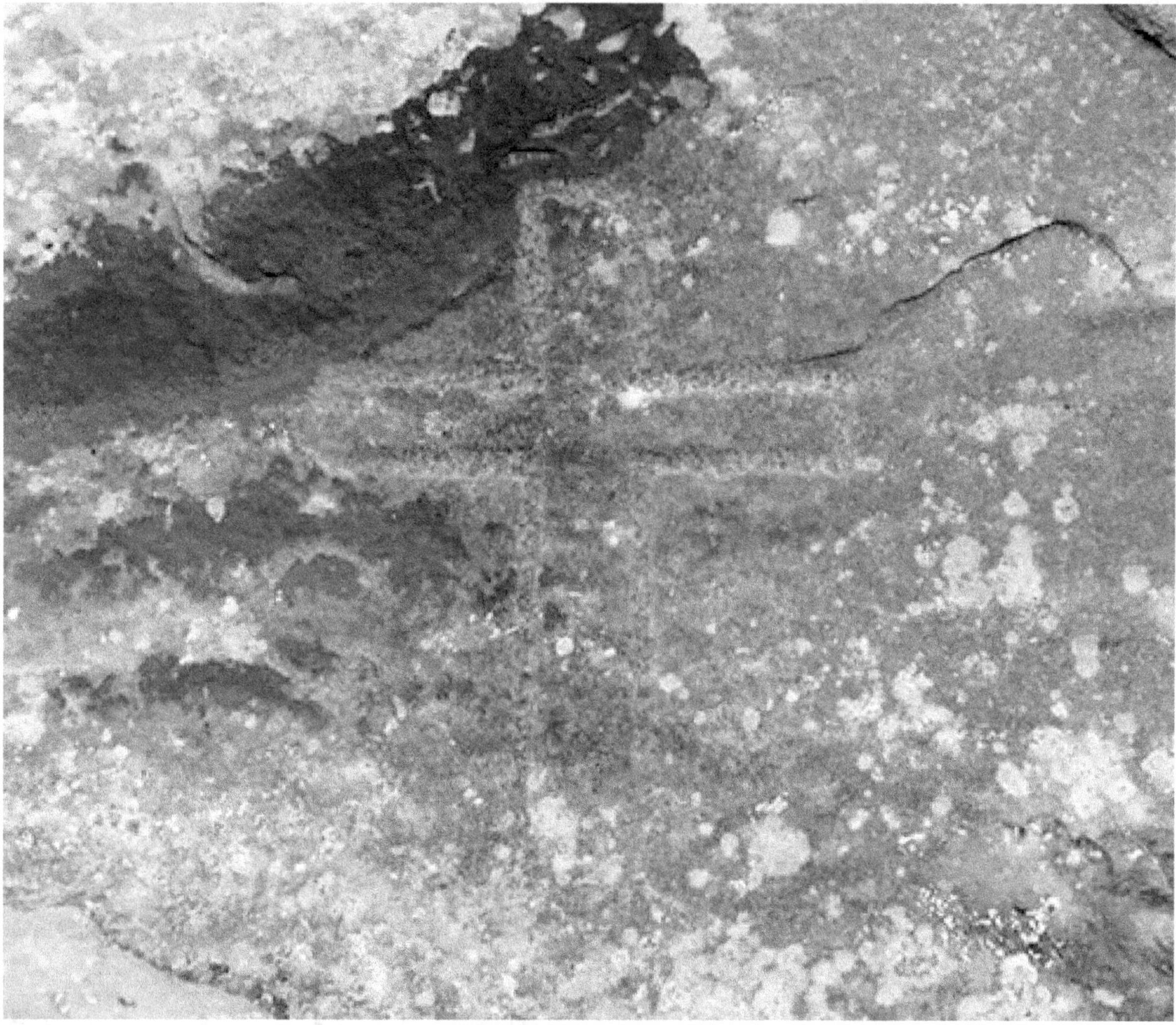

I call this one Ivie's cross because it's on the Ivie family's property. If you want to take a look, go to 40° 17.633′ N. 110° 29.116′ W., elevation 6282 feet. (You may need permission from the landowners.)

This cross and the next look as though they were chiseled or pecked into the rock. There is also a symbol to the left of the cross in the above photograph. I couldn't be certain what the symbol was or what it meant.

Please remember to enjoy this site as well as other locations without altering or disturbing them in any way. In viewing them, others will have the opportunity I have had to enjoy these precious pieces of our heritage.

This cross is in Diamond Fork Canyon. It's on the south side of the stream almost directly across the canyon from the sun symbol pictured on page 81. The cross can be found at 40° 04.773′ N. 111° 24.155′ W., elevation 6072 feet.

The Spanish have left many expedition markers. I have usually found them on trees. This time June found one chipped into a rock face. It's easy to draw comparisons between this one and those on Mosby Mountain and the one on the Daniel's Canyon map (page 205). It would be fascinating to pursue the research and determine which symbol belonged to which Spanish expedition group.

This picture is courtesy of June Balaigh.

More Rock Monuments

This row of rock monuments located in central Utah would have taken a fair amount of time to construct. I was impressed with the construction of these, and the fact that they are still standing. This picture is courtesy of Randy Lewis.

This monument is found in the same area as the one on the previous page. The picture is coutesy of Randy Lewis.

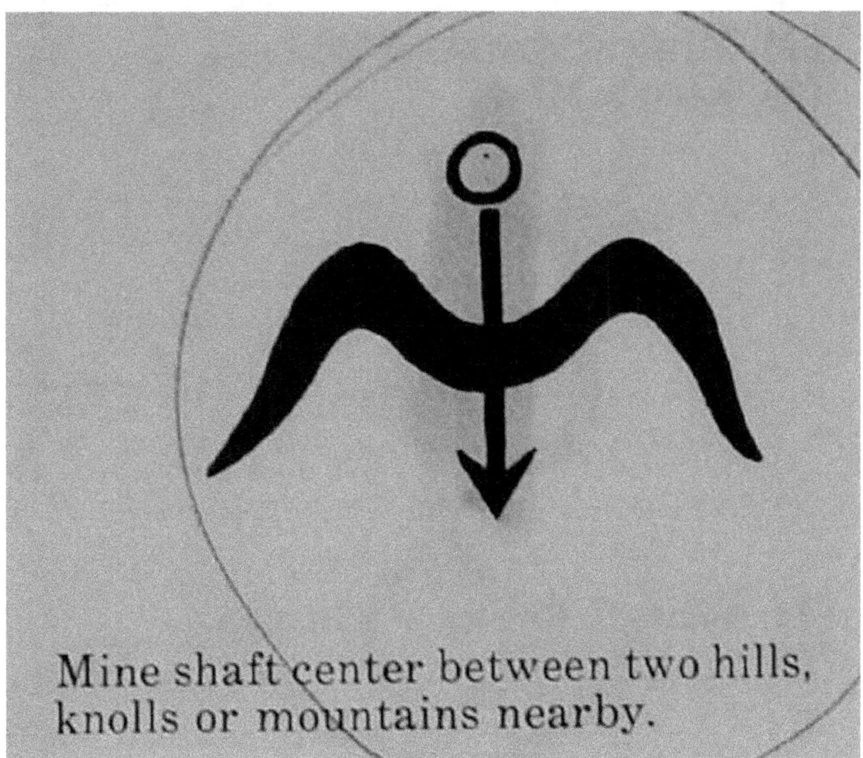

The rock formation above can be interpreted using this Spanish symbol.

The Upper Reaches of Whiterocks Canyon

These are symbols we found about 1986 while Bob Fair and I were traveling along a faint road on our way to Cliff Lake. The arrow at the bottom is pointing east toward the south side of Cliff Lake. The four circles could mean a distance of four varas (a Spanish unit of measurement equal to thiry-two to forty-three inches) or may have another meaning.

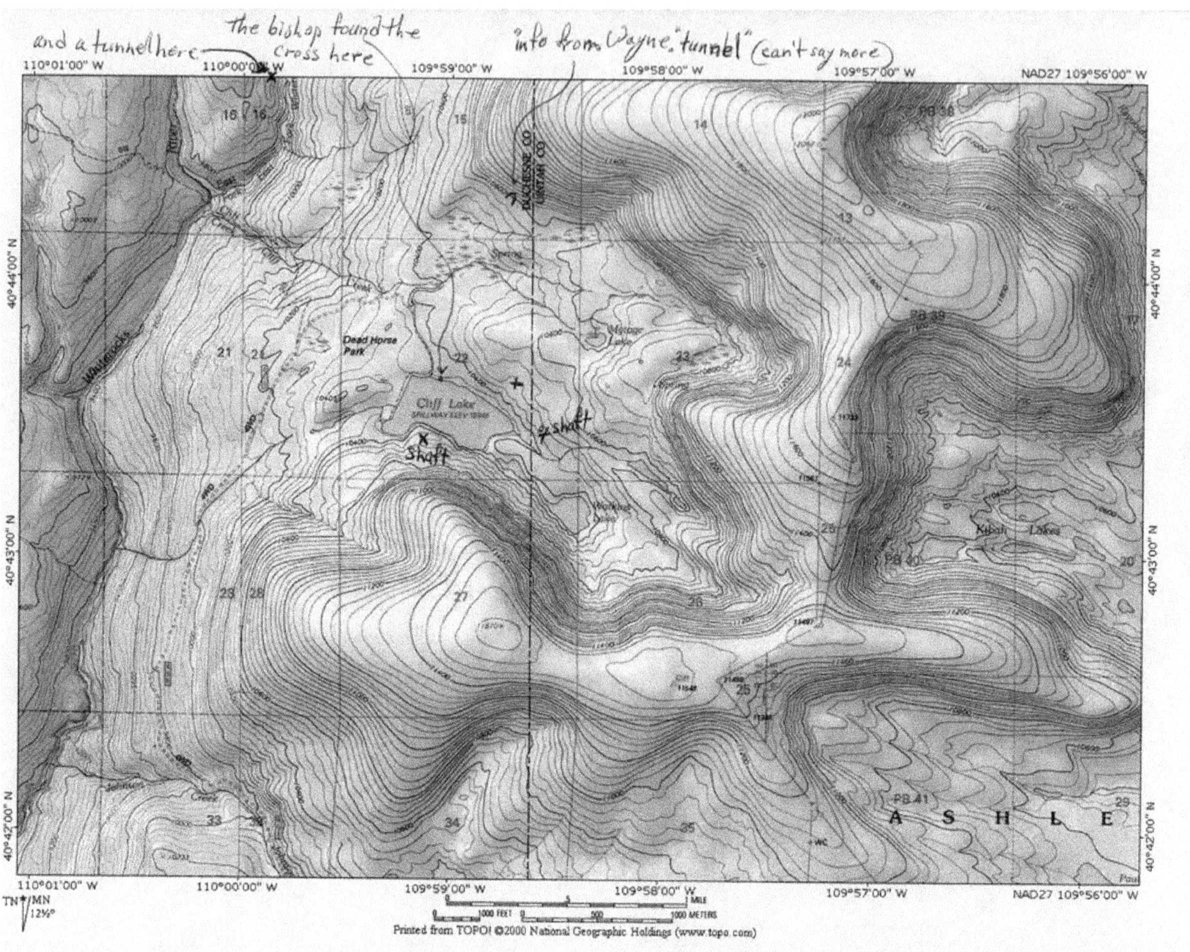

On this topographical map of Cliff Lake I have marked evidence of Spanish activity. The bishop found a Spanish silver cross near Cliff Lake. Map created with TOPO! © National Geographic Maps.

In 1972 while working for an oil company, a man named Richard found a skeleton in a cave low on the hillside just southwest of Watkins Lake. He took the helmet from the skeleton and hid it along with other articles, near Mytoge Lake for later retrieval. He came back the next year but had transportation difficulties and had to be lifted out by helicopter. He never had the chance to retrieve the artifacts. He gave me several clues, and when Bob Fair and I hiked in, luck was not with us, for we came out empty-handed. I hope someday that someone will retrieve these artifacts and place them in a museum.

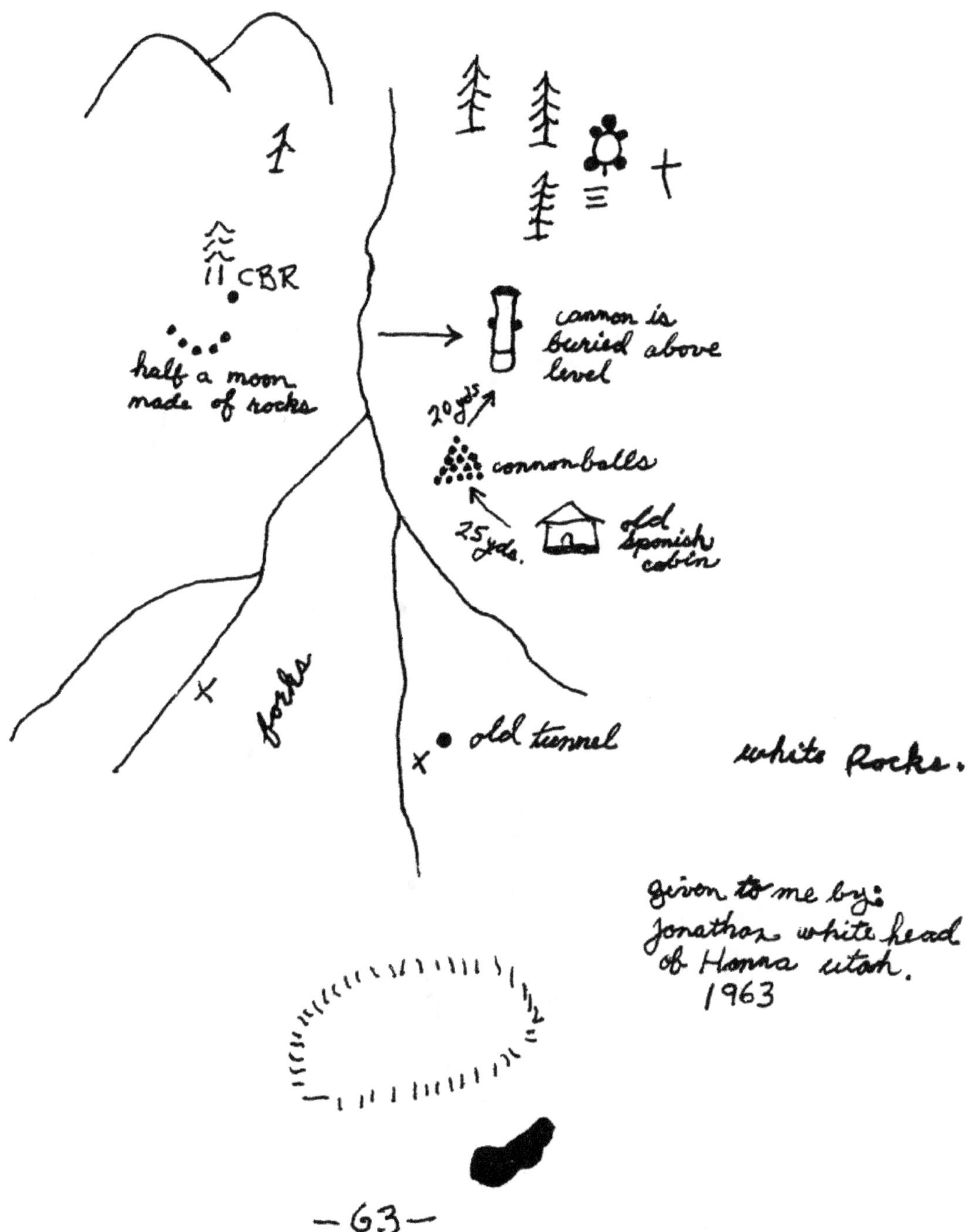

This is a map of the Whiterocks basin. As you can see, it needs to be turned around to make north the top of the map. Cris and I obtained it from the bishop and he obtained it from Jonathan White Head of Hanna, Utah, in 1963.

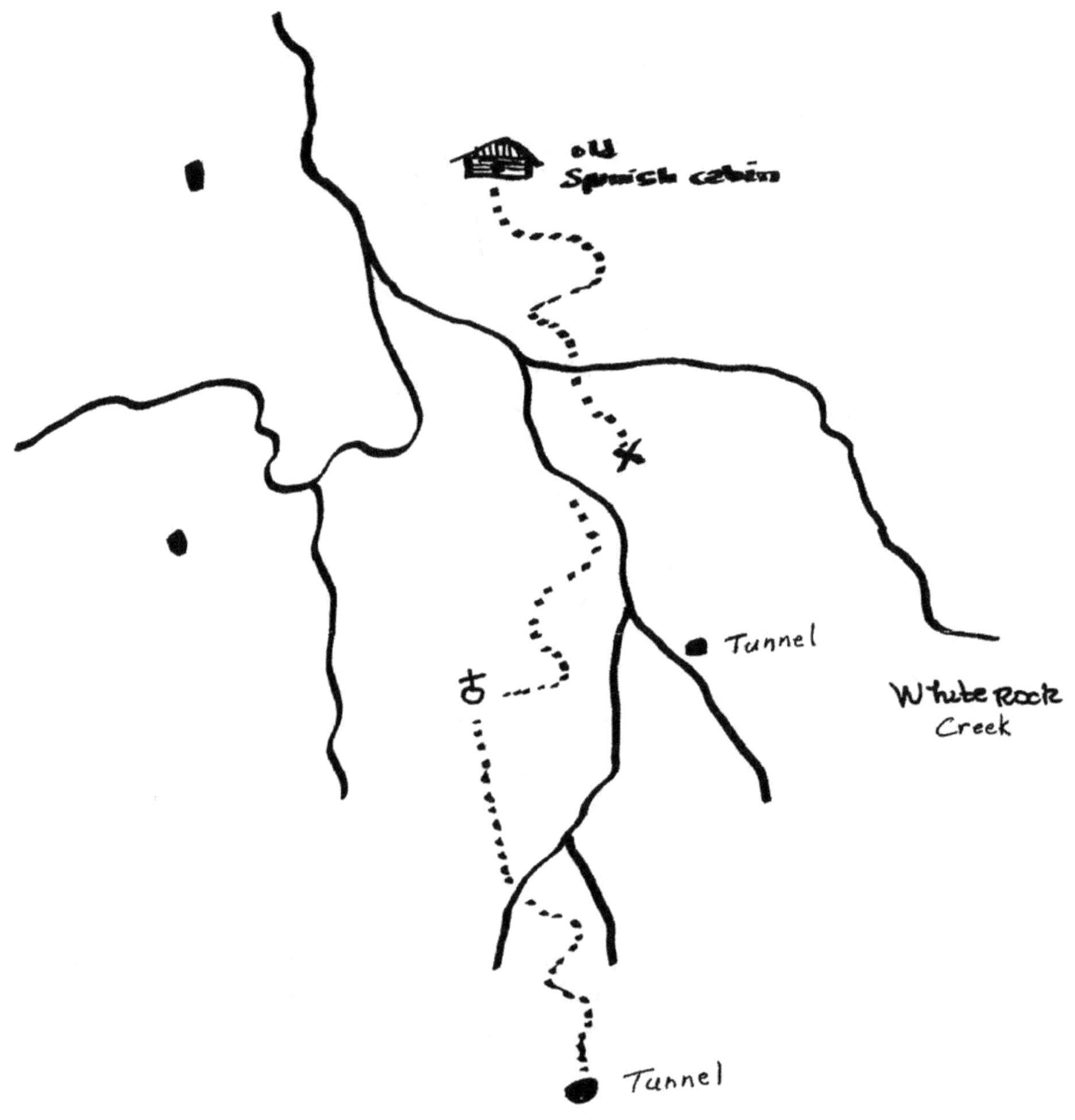

second version
of white head
map. 1963.

This is Aaron's version of the White Head map showing the upper reaches of the Whiterocks River. Notice one must turn the map around to make it match the canyon and the streams from north to south.

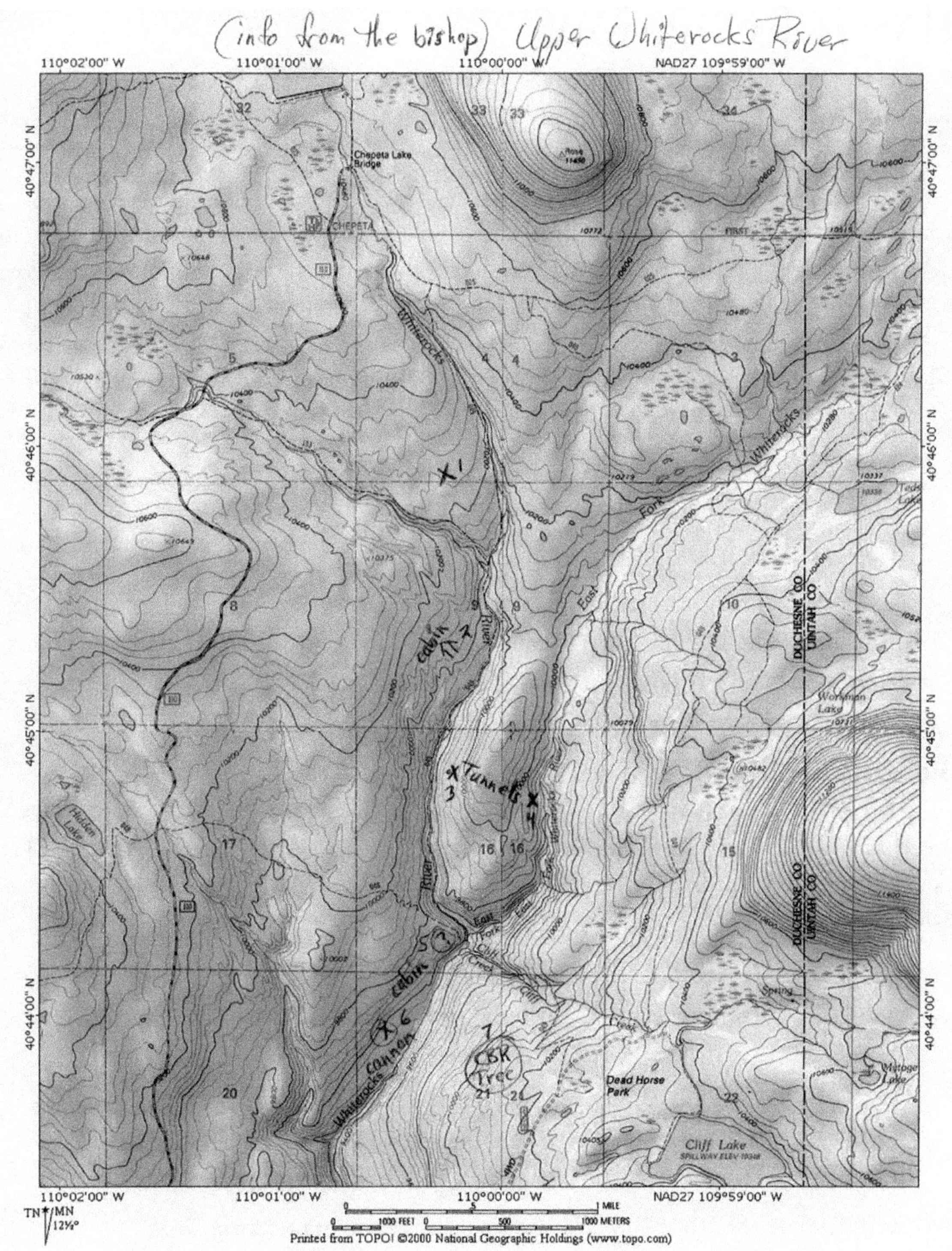

The streams match fairly closely once the hand-drawn map is turned around. I have approximated the location of the items shown on the hand-drawn map onto this map. On the next page are the approximate GPS coordinates of these items. This is work I have done and if there is an error, it is of my making. There are other areas where the bishop actually placed marks on a map, but this isn't one of them. Map created with TOPO! © National Geographic Maps.

GPS Information

Upper Whiterocks River

This is information I obtained from the bishop around 1995. These are estimated GPS coordinates from the Jonathan White Head map transferred onto a regular topographic map. These locations are not locations the bishop marked on my geographical map. I estimated where they might be. Therefore, if they are in error, it is my error and not the bishop's. This information is based on my best guess at trying to match up the White Head map with a topographical map!

These, and all other GPS coordinates in this book, are in the NAD 27 Alaska format. Some are listed in this book as degrees, minutes, and seconds, while others are listed as degrees and minutes.

1. 40° 45′ 54″ N. 110° 00′ 15″ W. Mine
2. 40° 45′ 17″ N. 110° 00′ 14″ W. Cabin Location
3. 40° 44′ 53″ N. 110° 00′ 13″ W. Tunnel
4. 40° 44′ 46″ N. 109° 59′ 52″ W. Tunnel
5. Disregard! I estimate the cabin is at location 2
6. 40° 43′ 57″ N. 110° 00′ 32″ W. Cannon
7. 40° 43′ 49″ N. 110° 00′ 06″ W. Area of a CBR tree

Other Locations in the Upper Reaches

The following are locations the bishop did mark on a geographical map for me. I have estimated as closely as possible where they are for GPS purposes. These aren't on a map within this book but are marked on my personal maps. These GPS locations could be transferred to your personal maps and be reasonably close, I believe.

8. 40° 41′ 31″ N. 110° 01′ 25″ W. Mine, below confluence with Johnson Creek
9. 40° 41′ 30″ N. 110° 01′ 10″ W. Shaft the bishop found in 1963
10. 40° 47′ 21″ N. 110° 01′ 28″ W. Tunnel, west of Chepeta Lake
11. 40° 50′ 05″ N. 110° 03′ 35″ W. Shaft, north of divide from Chepeta Lake.
12. 40° 50′ 19″ N. 110° 04′ 16″ W. Tunnel, north of divide from Chepeta Lake
13. 40° 45′ 56″ N. 109° 54′ 31″ W. Shaft, south of Deadman Lake
14. 40° 46′ 41″ N. 109° 54′ 37″ W. Vein and shaft, north of Deadman Lake
15. 40° 40′ 22″ N. 110° 06′ 06″ W. Old Ruins, west of Pole Creek Lake
16. 40° 39′ 53″ N. 110° 06′ 16″ W. Shaft, west of Pole Creek Lake

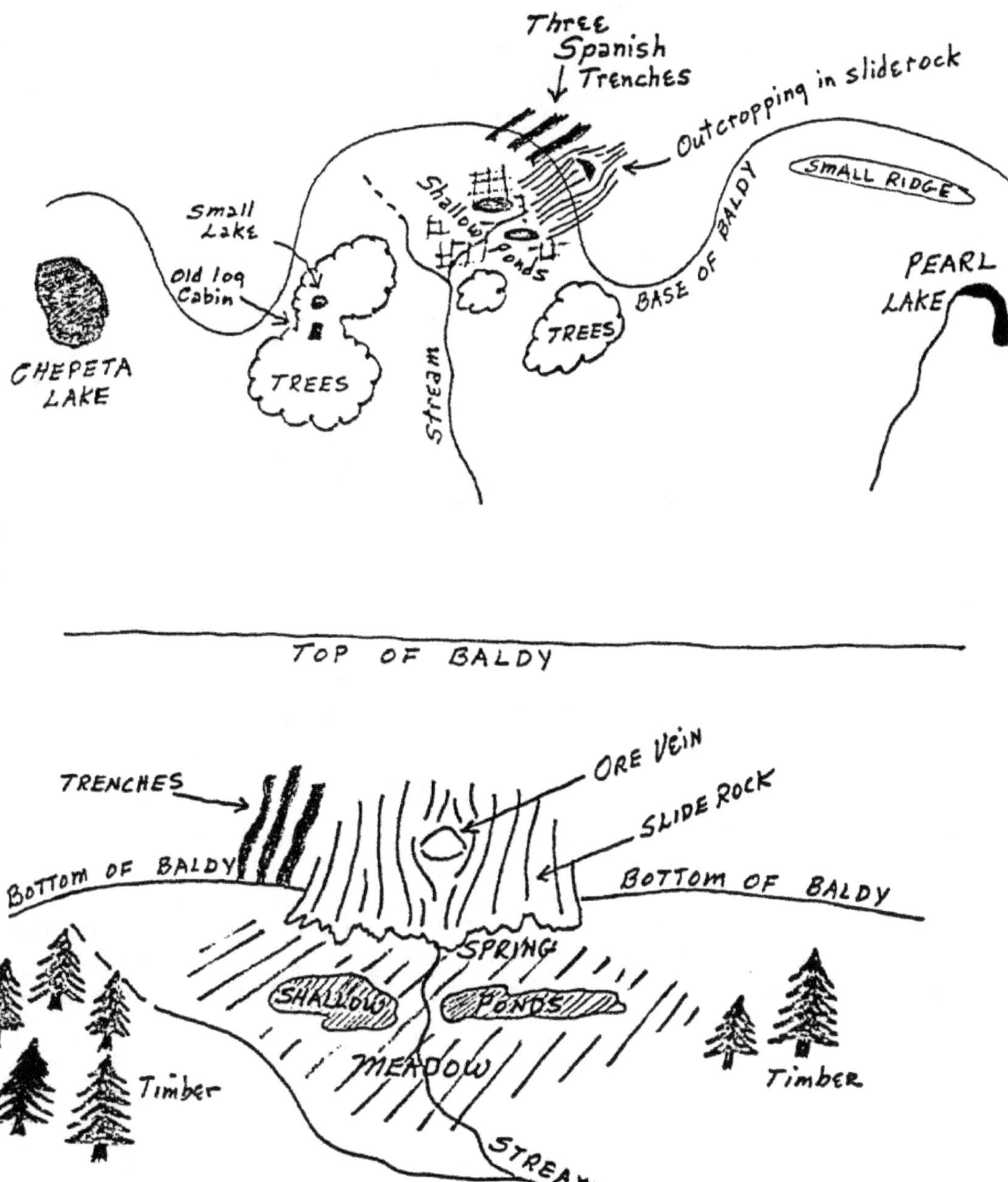

This appears to be the area Gale Rhoades and Kerry Boren were headed for on that fateful day when they lost their camping gear, equipment, and nearly their lives as they slid down a glacier while hiking in from the North Slope. You may read about it in Gale's book ***Lost Gold of the Uintah, the Rest of the Story***. He writes, "We failed to locate that rich vein of ore and to this day it still lies hidden in a contact zone beside an intrusive dike, covered by a few handfuls of slide rock." He goes on to describe the values as "96.3 ounces per ton in gold, 117.2 ounces per ton in silver and 20.3 percent copper." It just might be worthwhile to go and look for this one. Obviously Gale (aka Dusty) and Kerry thought it was.

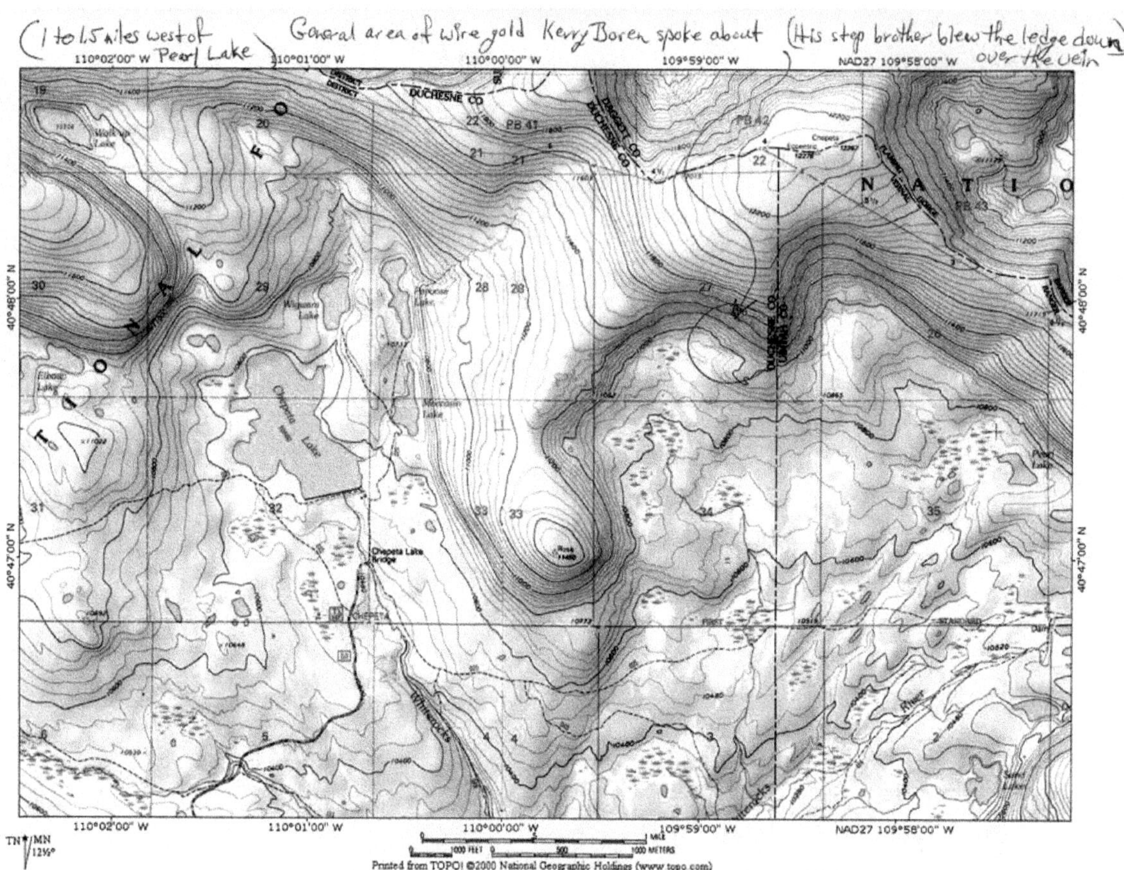

There is a certain spot with a discovery monument and digging northwest of Whiterocks Lake and Pearl Lake. Kerry Boren's family once pulled wire gold from it and staked a claim on this rich location. One day as they pulled gold from the lode and placed it into sacks, the sacks became so heavy they ended up leaving some of it behind to lighten their load. Kerry's step-brother tried to open up their digging with dynamite but only managed to bring the hillside down over this rich deposit. So rather than opening it up, he buried it. (This information is from personal interviews and correspondence with the Boren family.) Map created with TOPO! © National Geographic Maps.

NOTICE OF LOCATION

NOTICE IS HEREBY GIVEN, That the undersigned, having complied with the requirements of SECTION 2324 of the REVISED STATUTES of the United States, and the Local Laws, Customs and Regulations of this District, ha located Fifteen hundred feet in length by 600 feet in width, on this the All Valuable Minerals Lode, Vein or Deposit, bearing Gold, Silver, Copper, Lead and other valuable minerals, situated About 3 miles North west of Whiterocks Lake in Range 10E Township 2 North

in the .... Mining District, Uinta County, State of Utah, the location being described and marked on the ground as follows, to-wit:

Beginning 300 feet Northeastly of this location (Discovery) Monument, at the North end center monument, and running

thence South east 300 feet to S. E corner monument No. 1;
thence South west 1500 feet to S. W. corner monument No. 2;
thence North west 300 feet to N. W end center monument;
thence North west 300 feet to N. W corner monument No. 3;
thence North east 1500 feet to N. E corner monument No. 4;
thence South East 300 feet to place of beginning; including all Dips, Spurs, Angles and Variations

The above described Mining Claim shall be known as the White Rock

Date of discovery ....

Located this 1st day of July, 1956

DIAGRAM OF CLAIM

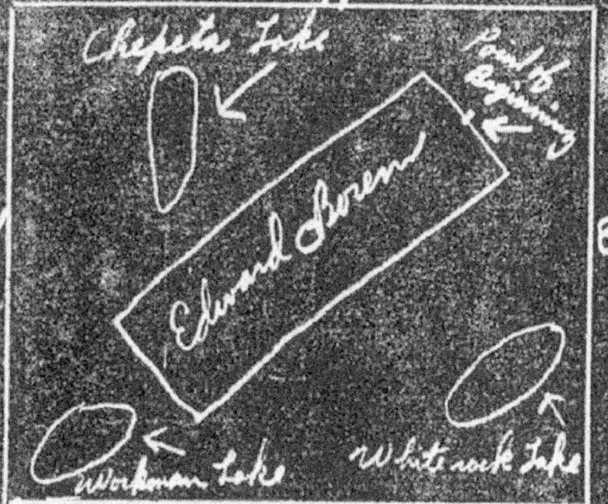

Chepeta Lake
Point of Beginning
Edward Boren
W
E
Workman Lake
Whiterock Lake

NAME OF LOCATORS:

Edward Boren

Entry 72128 5
Recorded at Request of Edward Boren July 25, 1956
at 4:00 P.M. Fee Paid $1.00 [signature] Uintah County Recorder

This is a copy of the actual mining claim staked by the Boren family.

My Experiences on Mosby Mountain

This photo of the smelter was taken looking down from the road.

On your way to Mosby Mountain take a look at the smelter below the road if you care to view a piece of our history. Some people say all the smelters were for the processing of lime. However, I have seen a glob of gold taken from the bottom of one of the smelters at Lime Kiln Springs. It was about one and one half inches long and three-eighths of an inch thick. That isn't a bad sized piece of gold for a lime kiln. The kiln shown here is on Indian land and is near a couple of silver mines. What would you guess it was used for? This smelter is along Little Water Creek.

Just north of the ice cave on Ice Cave Peak is where our story begins. In about 1981, my dad and I first visited this area and were intrigued by its mysteries. At that time there were so many trees decked with symbols that the task of following them was easy. Beckoning us on was a cross on a tree at a stone's throw north of the Ice Cave.

We continued north along this old Spanish trail. We arrived at Ice Cave Spring and kept going, all the while enjoying an abundance of symbols. According to Gale Rhoades in *Waybill to Lost Spanish Mines and Treasure,* the more flower petals there are on symbols like these, the richer the mine.

Hearts and diamonds were very common. An axe mark with a spike driven into it is said to indicate the direction to a mine. We found two trees in this area that fit that description.

The upper photos show PS and SP. Were these initials? The symbols in the lower photos represent Spanish expeditions. Gale Rhoades said there were as many as six individual expeditions working gold and silver mines in the vicinity of Mosby Mountain.

Along the way we saw many double S's. We also encountered initials, names, and symbols. Some symbols were very elaborate. Gale Rhoades said the double S's described a way around the cliffs to reach the mines. As we explored, we determined his concept was correct. When we reached the north end or top of the S's, the trees had been cut and large saplings had taken their place which made it virtually impossible to locate the mines on the north end of the S's.

A five direction tree was located and needs to be revisited with time to follow out the last or southern direction. This tree had five slashes on it cut at various heights and different lengths. It was as good as a roadmap.

One must also wonder who PS, SP, and DP were. Very few complete names were found on Mosby Mountain, but some have been discovered in other locations such as Rock Creek and Dry Fork.

Continuing north from Ice Cave Spring past many symbols, we came to this expedition symbol where the trail forked. (My dad is pictured here next to the expedition tree.) One trail led north to the mines located at the top of the Ss while the other leg led west and down over the edge into Whiterocks Canyon to the mines there.

The last time I saw this tree, or at least the section with the symbol, it was in an office in the Roosevelt Zion's Bank. It had been cut into two short sections, and one was stacked on top of the other. It saddened me greatly to see what had happened to one of my favorite trees.

We understand the heart at the symbol's top to represent the King's Mine and the diamond at its base to mean one league.

At this point we went west to the edge of the hill where we encountered two trees with symbols on them. They are pictured on the next page. The debarked box on the left tree had a large A cut into the upper left corner. There was a large heart in the box's center and a smaller heart in the lower right corner. Again, the large heart indicated the King's Mine or a rich mine. The symbols on the tree beside it consisted of an expedition symbol at the top, three diamonds below it, and then a cross at the bottom.

My dad and I had difficulty deciphering these symbols while simultaneously my former college professor became interested in this intriguing history. Bob Fair was his name and he stepped in to assist me with the deciphering of these symbols.

When I returned years later, both trees had been cut down, but not for their lumber; you see, someone had only taken the symbol section of these trees. The photos above are the before and after pictures of these important trees.

Bob Fair and the Crack Mine

In 1982, I enlisted the help of my former college professor, Bob Fair. He too enjoys a mystery, and it didn't take much to get him hooked on this one.

The symbols intrigued and perplexed us. We spent much time studying and interpreting them. We started at the ice cave just as my dad and I had, drove north to Ice Cave Spring, and then traveled on foot from there, continuing north along the old, but easily followed, blazed trail. When we arrived at the large expedition tree, we turned to the left, or west, and hiked out to the edge of the hill where we came to the trees pictured on the last page.

Hoping we understood the symbols in the box correctly, we placed our backs directly against the box on the tree, stuck out a left arm like we were pointing, and traveled in that direction down the hill. As we did so, we seemed to be following a faint trail which led us to the oldest cat face I had seen up to that point in my exploring adventures. This cat face tree stood at the base of a huge mass of nearly solid iron. If this indeed were the mine's location, then where was the entrance? Could it be sealed? Were we on the right track? We explored the area exhaustively. The terraced dips and mounds below us appeared as if they could have been natural except in a couple of locations.

The Uintas are strange mountains. Many times I have seen hillsides that have slid. However, here were places where the ground looked like it was trying hard to hide something . . . mounds drawn out directly from the mountain like a dump and rocks on the surface which were cracked and jagged, not smooth like those from a naturally occurring movement. The clues gave us reason to believe that mines were worked here long ago.

Bob staked a mining claim on the location and we started excavating. The enormous mass of iron looked promising and we chose to dig beside it.

All of our digging was by hand. It was hard work, but because of the steep and covered terrain, it was our best option. We started our dig at the base of the cat face tree and went north from there. Our dig was mostly uneventful, but at one point Bob got excited because he thought he had found a speck of gold on one of the chunks of iron. Unfortunately, it slipped off the rock he found it on and I never got to see it. I did, however, get my turn at excitement as we reached the north end of our excavation. I found something quite unexpected—a stick of dynamite. It was just as you might imagine, wet, soggy, and most likely highly unstable. Yes, that's right. I actually uncovered a stick of dynamite. This means someone had placed it there. Whether it was a botched attempt at blasting the mine's entrance shut or it had just been stored eighteen inches underground, we may never know. In any case, we were faced with the dilemma of what to do with it. I made the decision to scoop it up with my shovel, carry it to a new location, and bury it. I gently set it down on the lower edge of our newly formed dump, dug a new hole, and gently set it inside. It was quickly buried, but not soon forgotten.

Here is the oldest cat face and the crack mine as we came to call it. In the lower photograph, just beyond the iron boulder, the dynamite stick was uncovered and then relocated and buried deep in our tailings pile.

We continued to excavate a little more, but our efforts seemed to be in vain. It was difficult to determine which way the mine entered the mountain, if indeed there had been a mine. We looked it over and looked it over again. Indeed, we were perplexed and finally came to the decision to halt our efforts and call it quits. It had been an interesting adventure. We were none the worse for wear and had enjoyed the experience.

Bob, however, couldn't leave it alone. He hired a man with a Caterpillar tractor to go in and continue to excavate where we had stopped. The man Bob hired made it to the edge of the hill just south of the ledges and north of our dig, and started down the steep terrain when he ran into problems. He was unable to make it to the dig for one reason or another. I was never told exactly why and I wasn't there at the time.

In any case, he had to go back and bring in a larger Cat to pull the first one out. He was able to do so, but in the process the Forest Service caught up with Bob and the excavator and told them in no uncertain terms that there had not been a "Plan of Action" filed with the Forest office and they were to stop this type of work or legal action would be taken against them. At this time, Bob had had enough and gave up his efforts to open the mine at the base of the iron mass.

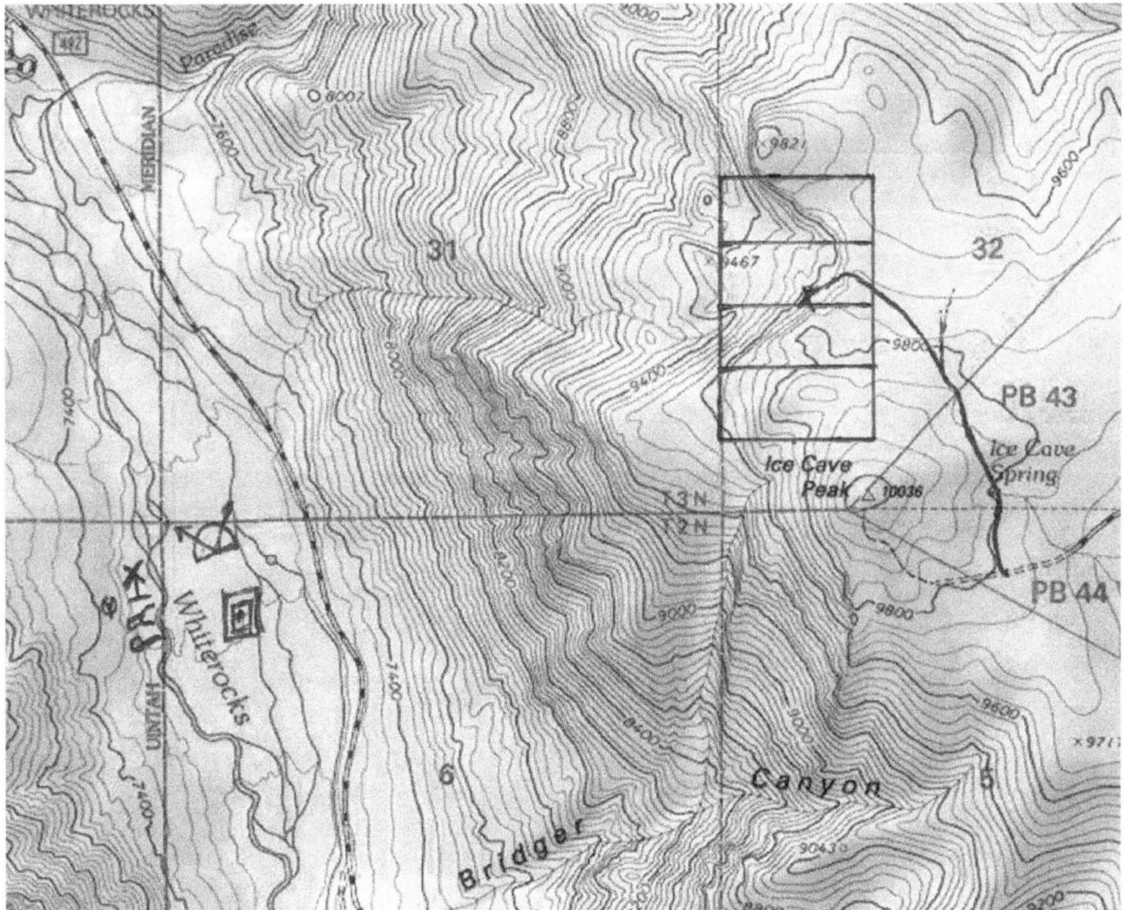

This map indicates our claims and mining adventure from the early 1980s. Map created with TOPO! © National Geographic Maps.

Bob's next mining interest was at Mosby Park. He noticed the claims on the Iron Mask Mine hadn't been renewed and he filed over the Atwood claims. I was a silent partner by this time, which was fine with me. Walt Farmer was the financial backer with Kelly Healy as the assayer and field geologist.

The Iron Mask Mine sits at the west end of Mosby Park at the base of a large hill. There is an iron blowout about two hundred yards above it with iron and large chunks of calcite around the hole. The red and cream-colored clay attached to the iron at the blowout seemed to carry the highest concentrations of gold and silver of any of the sampled rock. The mine itself had an iron door on it when I first saw it. Whoever had dug the tunnel, it seemed, had attempted to intersect the iron core of the blowout but was off a little to the south. Or at least that's how it looked to me.

Hundreds of samples were taken and analyzed. A formal proposal was drawn up and offered to a Canadian mining company. They considered it but turned it down. We didn't have enough backing to attempt the project ourselves. So once again we gave up our dreams.

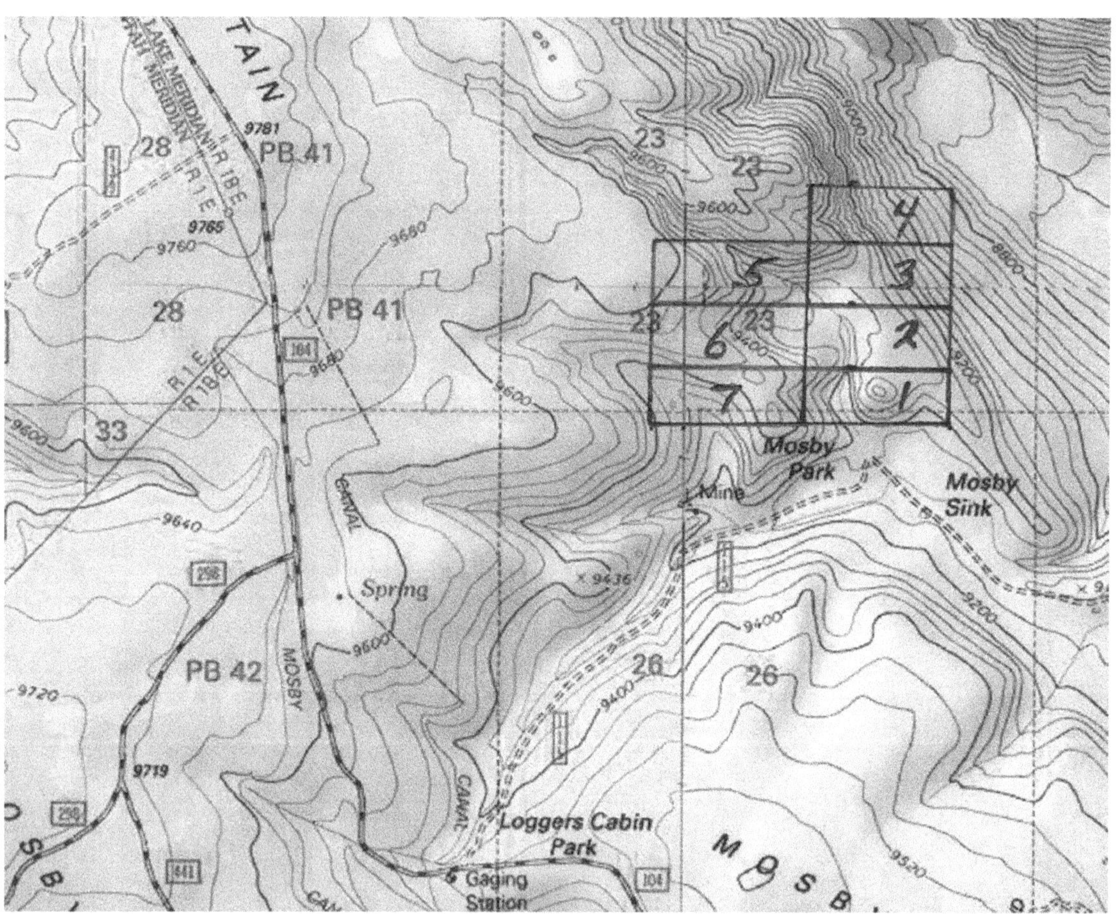

This is a map of the claims at Mosby Park. Map created with TOPO! © National Geographic Maps.

This is the portion of the mine that I dug.

In the late '80s, as Bob was engaged with the claims near the iron mask mine, I tried my hand at opening a mine that I had found. It was between Logger's Cabin Park and Mosby Park about one quarter of a mile southwest of Mosby Park. It had caved in but held promise to an inquisitive mind like mine. One thing hampered me: I didn't know the background of the mine. I was alone in this project, but I liked that aspect because I could call the shots. The hillside above the mine had collapsed into it. Because I thought cleaning out the entire mine would be a huge undertaking, I decided to go up the hill and dig straight down into the back of the mine behind the blockage. Occasionally family members and friends helped. As the hole got deeper and deeper, it became apparent that the sloughing hillside above the hole was increasing in size and becoming more and more dangerous. We would have to shore up the hill to continue. That reality, and the fact that I'd found only iron and quartz up to that point, helped me with my decision to let someone else clean out this mine.

Here are the GPS coordinates for the mine: 40° 37.063' N. 109° 52.529' W., elevation 9287 feet. If you don't have a GPS unit, you may travel from Logger's Cabin Park toward Mosby Park. About one-quarter mile before arriving at Mosby Park, look for a cat face tree to your left. The mine is about one hundred yards behind this cat face symbol. The old tunnel entrance is downhill from this hole about twenty feet. Good luck to who ever decides to tackle the project. I hope it's a rich one. The portion I dug is pictured below.

The Sighting Hole

One beautiful, sunny day in the spring of 2002, Randy Lewis, Ken Lance, Jerry Ann Lance, Mick Norton, and I consolidated our efforts in locating a mine. We had bumped into each other quite by chance that day on Mosby Mountain and decided we had a golden opportunity before us. Mick wasn't able to hike but was eager to share some information with us. We drove to a pretty meadow, hopped out of the vehicles, and anxiously awaited our instructions. He showed us which hill to hike and told us what to anticipate. We hiked up the hill and located the cave in the cliff just as he had described it. As we crawled inside, we had to wonder if this was natural or manmade because of its size and shape. At the far end, we found the sighting hole just like Mick had described it. We looked through it, took our bearings, and did our best to memorize the landmarks. It wasn't hard because there was a large dead pine tree clearly visible through the hole. We now knew exactly where we were going and the anticipation ran high. Quickly we exited the cave, slid down the hillside, and headed across the meadow toward our noted landmark. We found the dead tree and only a short distance from it we found the mine Mick had been seeking. It had originally been a vertical shaft, but now it was only a large depression in the ground. If we could validate this mine's worth and determine the amount of effort necessary to reopen it, we could justify a mining claim and clean it out. Maybe we will hold off until we know more about it. However, if the Spanish miners went to that much work to mark this mine (making the cave and the hole), then it stands to reason it must be worth excavating.

This is my friend Randy Lewis standing in front of the cave in the cliff.

Now you see what we saw as we examined the hillside from within the cave.

This is a picture of the depression or mine we found after following the clues. If you feel inclined to take a look, you may start your search at Mosby Park as we did or go directly to this depression located at 40° 37.211'N. 109° 52.472'W., elevation 9303 feet. Good luck.

Moon Lake and the Copper Map

The story of Moon Lake and the copper map is an intriguing one. With the legends in mind and a sense of the ancient and modern history involved, I soon came to understand that this is a sacred spot and has been for many generations. Not that long ago, the Native American people held their sacred Moon Ceremony here. This is the basic story as it was conveyed to me.

It's a moon lit night and a small group of braves, elders, and a shaman is preparing for this sacred event. The ceremonial stone with five depressions is prepared. The four outer holes contain the gold from the sacred mine while the center hole is filled with water from the lake. Now they need to wait a short time until the moon's reflection can be seen in the pool of water. Soon their ceremony starts. It will continue until the business at hand is accomplished. When the ceremony concludes, the shaman retrieves the precious yellow metal and returns it to its rightful resting spot: the sacred mine not far from the water's edge. The surroundings are hard to miss. There's a pine tree with an eagle carved deep into its bark and a short distance away a rocky outcrop with the all-seeing eye engraved upon it.

It won't be long until the Spanish come and take this mine from them. In the meantime, they will honor their maker and continue their ceremonies. They will also share this precious metal with the members of The Church of Jesus Christ of Latter-day Saints, often called Mormons, led by Brigham Young. Thomas Rhoades will be sent for the gold until he is no longer able to go for health reasons; thereafter, his son Caleb will take his place. Legend further states that Chief Walker will ask an Native American named Ridley to escort Thomas and Caleb to this and possibly other mines. (This Ridley may have been an ancestor of Richard Ridley, who was very good friends with Ed Twitchell. No wonder Ed knew so much. See the chapter devoted to Ed.)

Thomas Rhoades did come. He was allowed to be a protector and courier for the Church's monetary needs by bringing gold to Brigham Young from this mine. Brigham Young commissioned James M. Barlow to make a copper map as a permanent record of this event and of the mine's location. That was in 1863.

Caleb followed in his father's steps by going to the sacred mine at Moon Lake. Caleb found other ore-bearing locations not far away and marked those as his claims by carving his initials into nearby trees. Caleb never divulged the location of the sacred mine. His father, Thomas, was also a man of integrity and kept the location of this sacred mine secret except to the Church, and that record is permanent on a thin sheet of copper.

It may take a lifetime to put all of the puzzle pieces into place and see all that went on here, but that's part of the fun and the thrill of the hunt. Moon Lake truly is a sacred and a remarkable location. It's breathtaking in its beauty and incredibly intriguing with its mysteries.

A Few Details about the Copper Map

Most likely, the copper map once belonged to The Church of Jesus Christ of Latter-day Saints, also known as the LDS Church. I have reason to believe the church had possession of it until about 1901 when it was checked out. It remained with that individual and his descendants until 2000 when I gained access to it. It measures eleven inches by nine inches and is very easily marked or scarred.

Because it is a piece of our history that could link the LDS Church with the Rhoades Mines, I have chosen to share it in part with you. Before doing so, the Church was contacted. An email was sent to the LDS Church on March 31, 2006, requesting permission to disclose this map in a forthcoming book. Here is their reply dated 17 April 2006: "After reviewing your request, we have determined that we are not able to grant permission to use this map as our records do not indicate if this map was ever in Church Archives. Thank you for checking with us."

The decision at that point was left up to me. Even though I can't technically link this map to the Church, I still believe it is authentic. I have decided that, because of the sacred nature of this map and geographical area, I will not disclose its complete detailed image at this time. I have chosen to keep this map small and blurred, and possibly avoid a stampede, while giving expanded views of certain areas of this map.

An image of the entire copper map is found in the center left of the next page. I have expanded three areas for further discussion. The upper left portion of the map is its title. It reads:

DESERET ASSAY OFFICE

G.S.L.C.D.A.O.

THOMAS RHOADES MAP

SECOND MINE NEAR RHOADES VALLEY

MOON LAKE DESERET TERRITORY

BY. JAMES M. BARLOW

1863

THE CHURCH OF JESUS CHRIST OF

LATTER DAY SAINTS

The symbols on this map can be seen in the other expanded areas and have been deciphered by Stan Johnson. He has studied the Egyptian language for more than ten years. His explanation begins on page 124. (Stan wrote the informative book *Translating the Anthon Transcript.*)

In order for these symbols to have been placed on the copper map in 1863, they must have been on a rock or other object in the vicinity of Moon Lake. We can conclude that Thomas Rhoades must have seen them and recorded them prior to that date. Gale shows some of these symbols in *Lost Gold of the Uintah, The Rest of the Story*. He says that Carlos Foote found them near Moon Lake. With luck and time we too may find them, take pictures, and document our findings.

The Copper Map

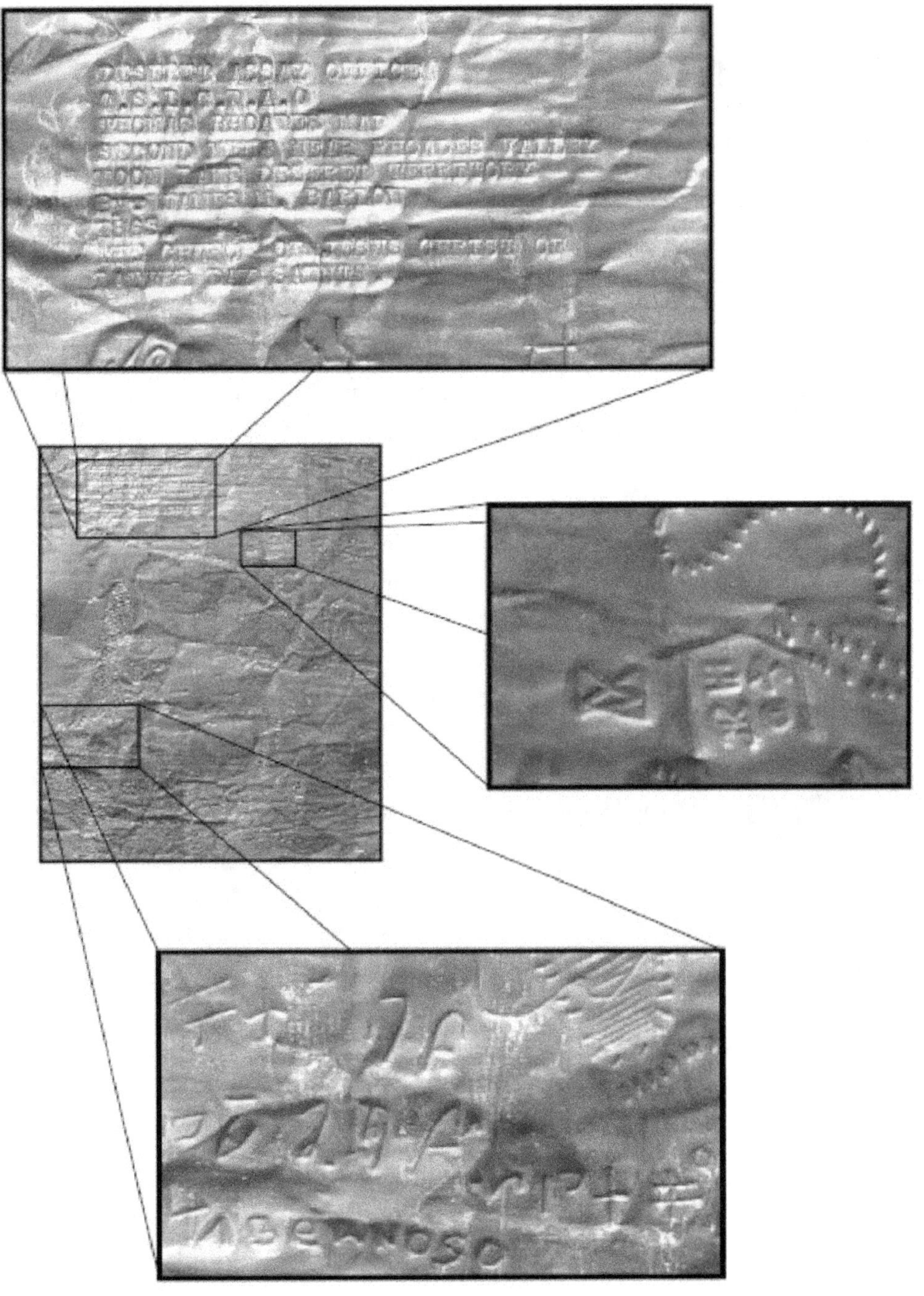

The Symbols on the Copper Map

These are the symbols on the copper map. According to author and scholar Stan Johnson, they are the best example originating in Utah of reformed Egyptian he has seen. He deciphered them for me. Here are the symbols and their meaning according to Stan:

The house symbol is exactly what it appears to be. Stan says we may actually be able to date these symbols using this one as a reference. This symbol was made after the white man's influence or it would be a different type of dwelling.

Wait until the right time. Then whatever is being held can be divided.

A reference to holding or having. The dissecting line means waiting with the possession.

Not or don't enter. The inward lines mean strong borders or impenetrable. The indentation at the top means receive or open.

A deep, covered chamber.

Opposition or war. It further denotes that those holding it will win if put to the test.

In reference to the past, meaning sacred

Coming forth

Sacred

Being held and positioned

Held in a chamber

Being held

The line above means covered. The balance means a deep covered chamber.

Going or walking

Restricted and positioned

The line in the middle means wait or held. The curve at the top denotes below or opening.

Opening

Sacred

Coming forth

Being held below the surface. The circle means sacred, pure metal or stone.

Tabernoso is Spanish. It literally means "looks like a tavern." We can extrapolate some and apply it to the surroundings to mean a dark, possibly cluttered shelter with many things lying around.

Placing the individual symbols together, it is read like this: "Something sacred is being held beneath, in a chamber, below is being held. Something will come forth from a chamber (like birth). That which is held will come forth."

It's interesting to note that these symbols form a chiasmus like the symbols which were translated from an ancient record of the early inhabitants of America and became known as the Book of Mormon. This is further verification that these symbols are reformed Egyptian.

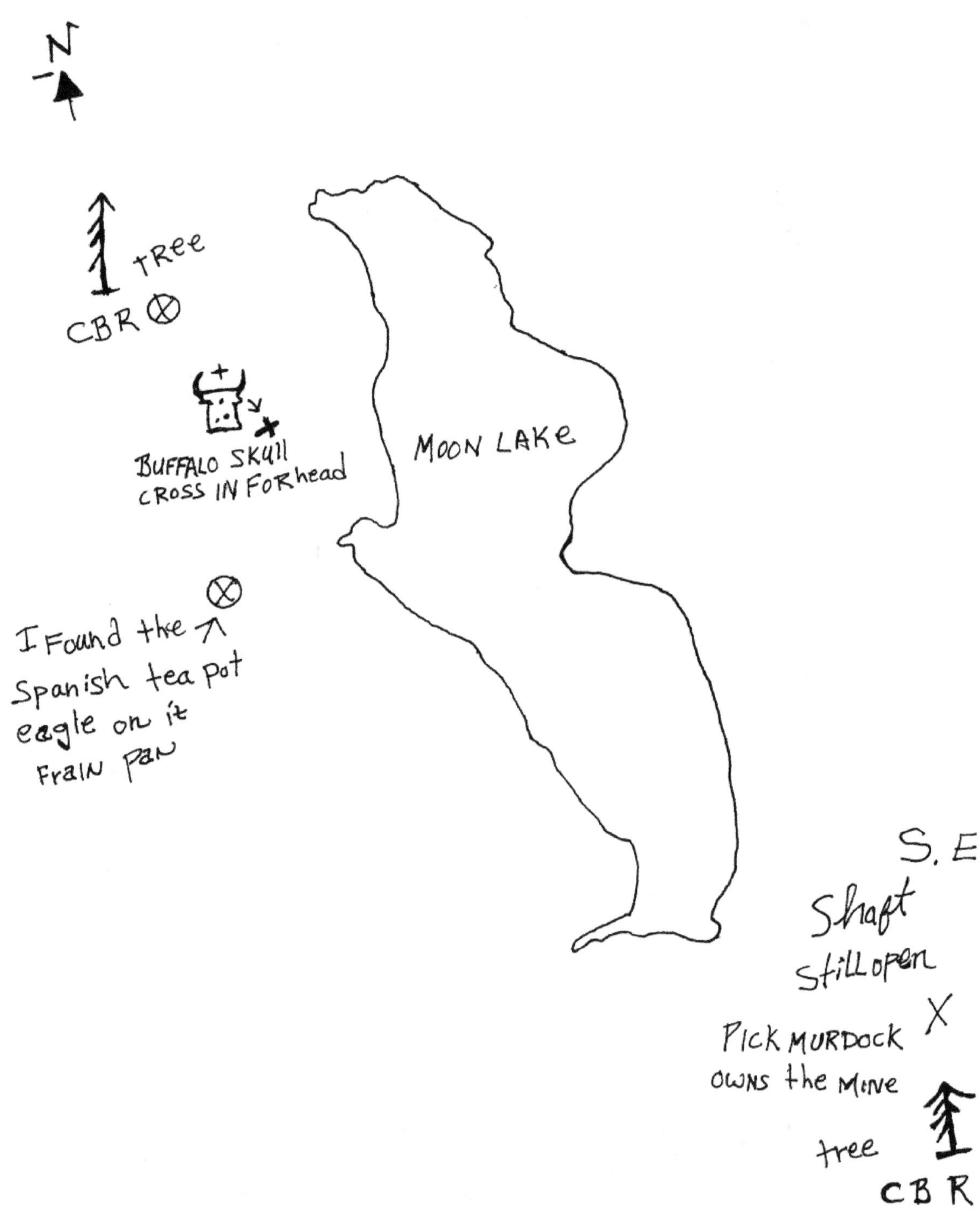

Aaron worked for the government at Moon Lake during its expansion and dam-building phase in 1935–38. He found several things, including mines, two trees with the letters CBR carved into them, a Spanish teapot, a buffalo skull with a cross carved into it, and much more. (It is believed that CBR carved into the bark of trees was the usual method Caleb Rhoades used to mark his mining claims.)

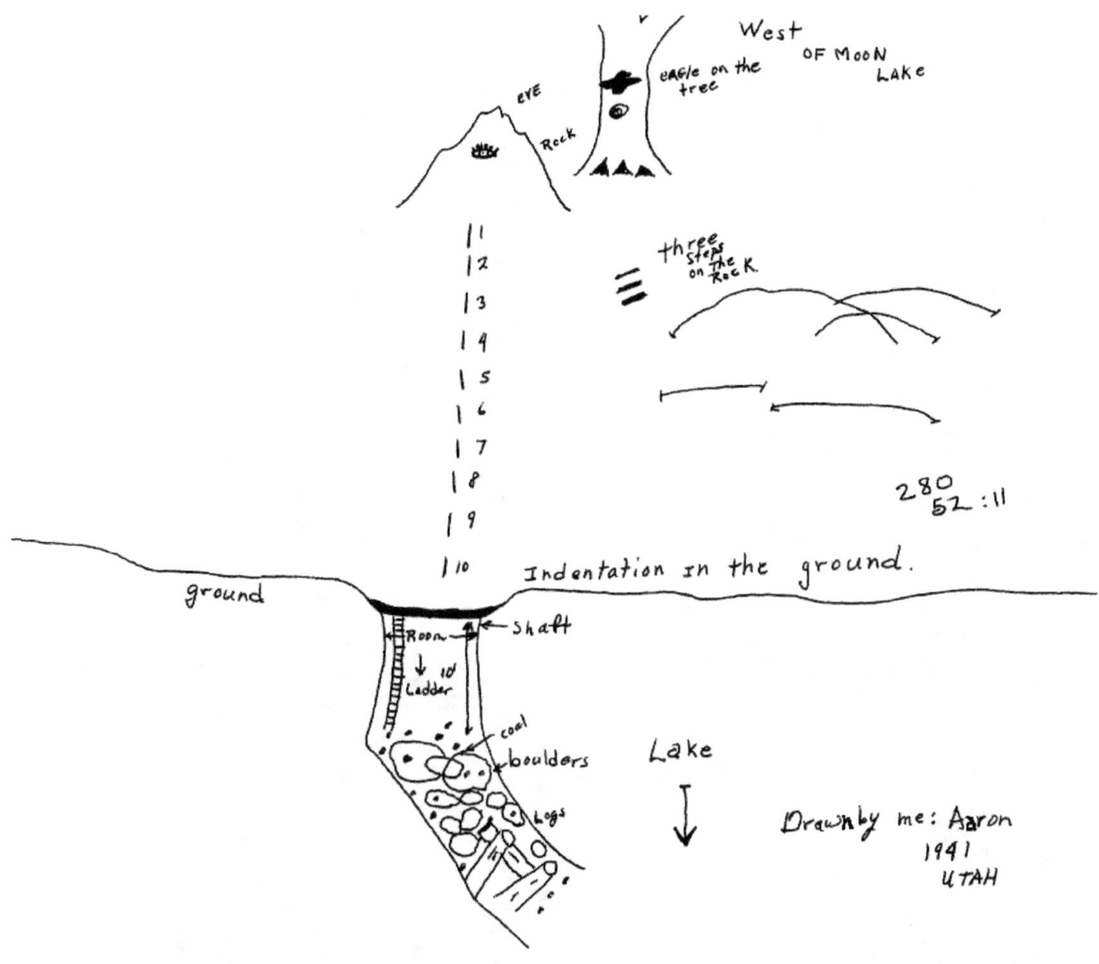

This is the depiction Aaron drew of the sacred Native American mine near Moon Lake. When he found this mine, there were horse droppings around the entrance as well as topaz crystals and small clusters of quartz crystals with wire gold visible in the quartz. He put his back to the rock marked with the all-seeing eye on it, and walked ten paces toward Moon Lake. He descended the ladder and made his way amongst the boulders, logs, and rocks that looked like coal to reach an elliptical ore body at the end of the tunnel. Here he chipped off a piece from the right-hand side of this ore body about half the size of a fist. The gold was highly visible, being about three-fourths of an inch thick along its edge.

Aaron asked the shaman to bring him enough gold from another mine in the area to make wedding bands for Aaron and his wife. The shaman went to a mine near Moon Lake for which he had the key to the outer and inner doors and retrieved enough gold to fulfill Aaron's request.

This story is like some other stories in this writing, in that Aaron and the bishop both knew about this mine. They were good friends and shared much information.

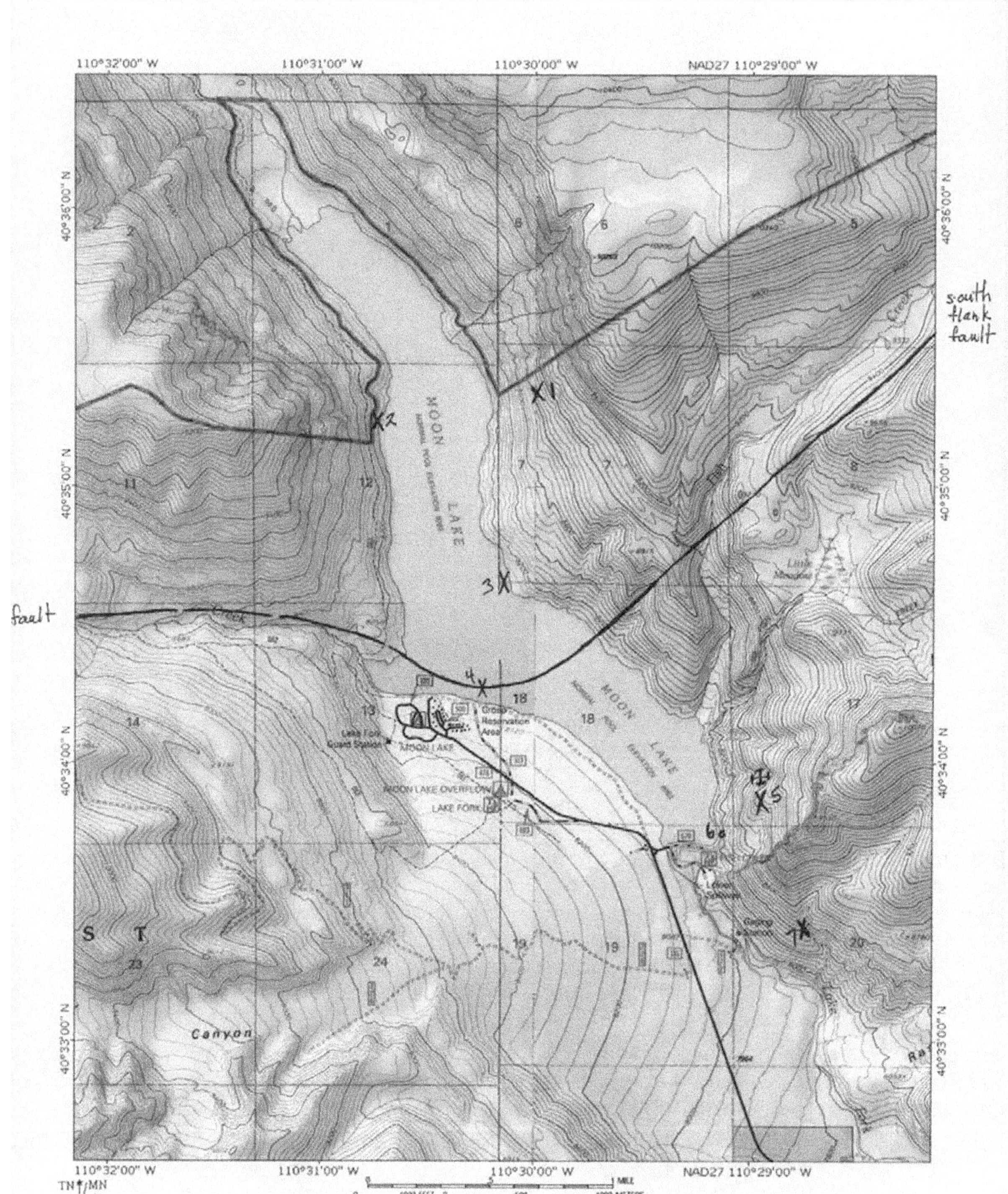

This map of Moon Lake shows the approximate location of the South Flank fault. It also shows dowsed locations that Pedro Angulo dowsed. On the next page are the approximate coordinates of these dowsed locations. Map created with TOPO! © National Geographic Maps.

GPS Information

Moon Lake

Dowsed by Pedro on October 16, 1995

Estimated coordinates in the NAD 27 format.
These are the approximate GPS coordinates of the locations Pedro dowsed.
(Refer to 1 through 7 on the corresponding map.)

1. 40° 35′ 20″ N. 110° 30′ 00″ W. Mine of Gold and Silver. Could be Very Rich
2. 40° 35′ 15″ N. 110° 30′ 43″ W. Gold, Very Close to the Water Line
3. 40° 34′ 40″ N. 110° 30′ 09″ W. Gold Mine in the Lake
4. 40° 34′ 16″ N. 110° 30′ 15″ W. Gold and Silver Mine in the Lake
5. 40° 33′ 52″ N. 110° 28′ 58″ W. Very Rich Gold Mine. Beware! Bones Inside
6. 40° 33′ 44″ N. 110° 29′ 09″ W. Small Amount of Gold
7. 40° 33′ 25″ N. 110° 28′ 46″ W. Rich Mine of Gold and Platinum

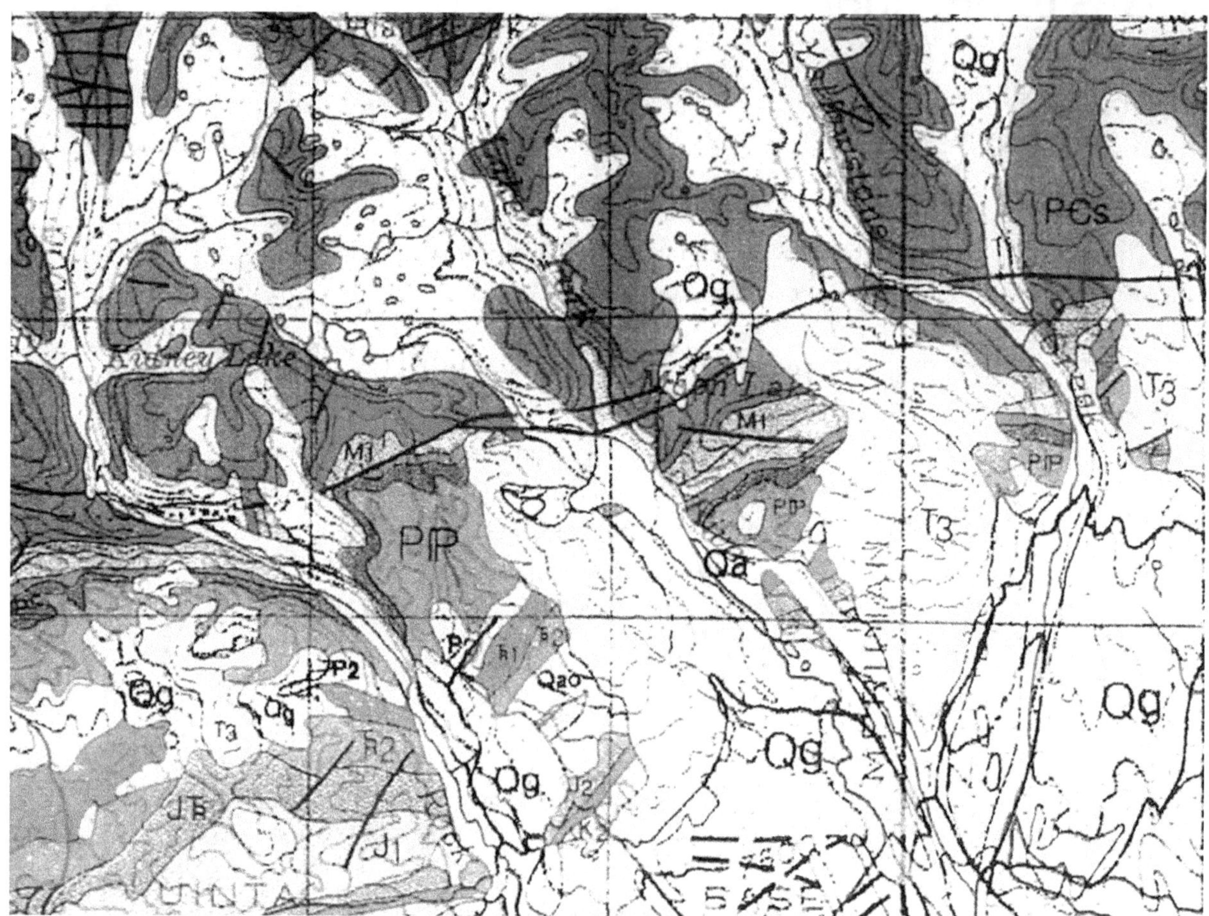

This is a geological map showing the faults that intersect Moon Lake. Moon Lake is in the center of this map.

The above photo shows a shaft that Aaron knew about but never investigated. It also points the way down the canyon to another mine that he described as being open and on a ridgetop. I believe it's on the ridge just south of Raspberry Draw, but that location hasn't been verified.

These cliffs near Moon Lake are captivating. Legend says there are several mines nearby.

Moon Lake, to me, means very old symbols, breathtaking scenery, and plenty of mystery.

With the use of a small core drilling tool, this symbol was determined to be about 240 years old. That would place its creation in about 1771—the date Nunez was exploring and mining in the Uinta Mountains. See the map on page 208.

Randy and I examined several circles of rocks and determined that they were placed there long ago. The lichen growth is a sure way of knowing.

A friend named Randy Lewis and I explored nearby and found four circles of rocks and several diggings.

Randy and I found three or four diggings like this one in a line leading us down near the shore line of Moon Lake.

Between the owl symbol and the shoreline, we found this unusual pile of rocks. It didn't look natural to us and we believe the rocks have been moved.

The upper pile of rocks has been disturbed. We also found iron nearby. The adventure continues.

Within a short distance was another outcropping that hadn't been disturbed. Here was a layer of iron up to one-quarter inch thick coating the rocks. The iron is a good sign and other minerals may be nearby. We'll keep looking for the precious yellow ones.

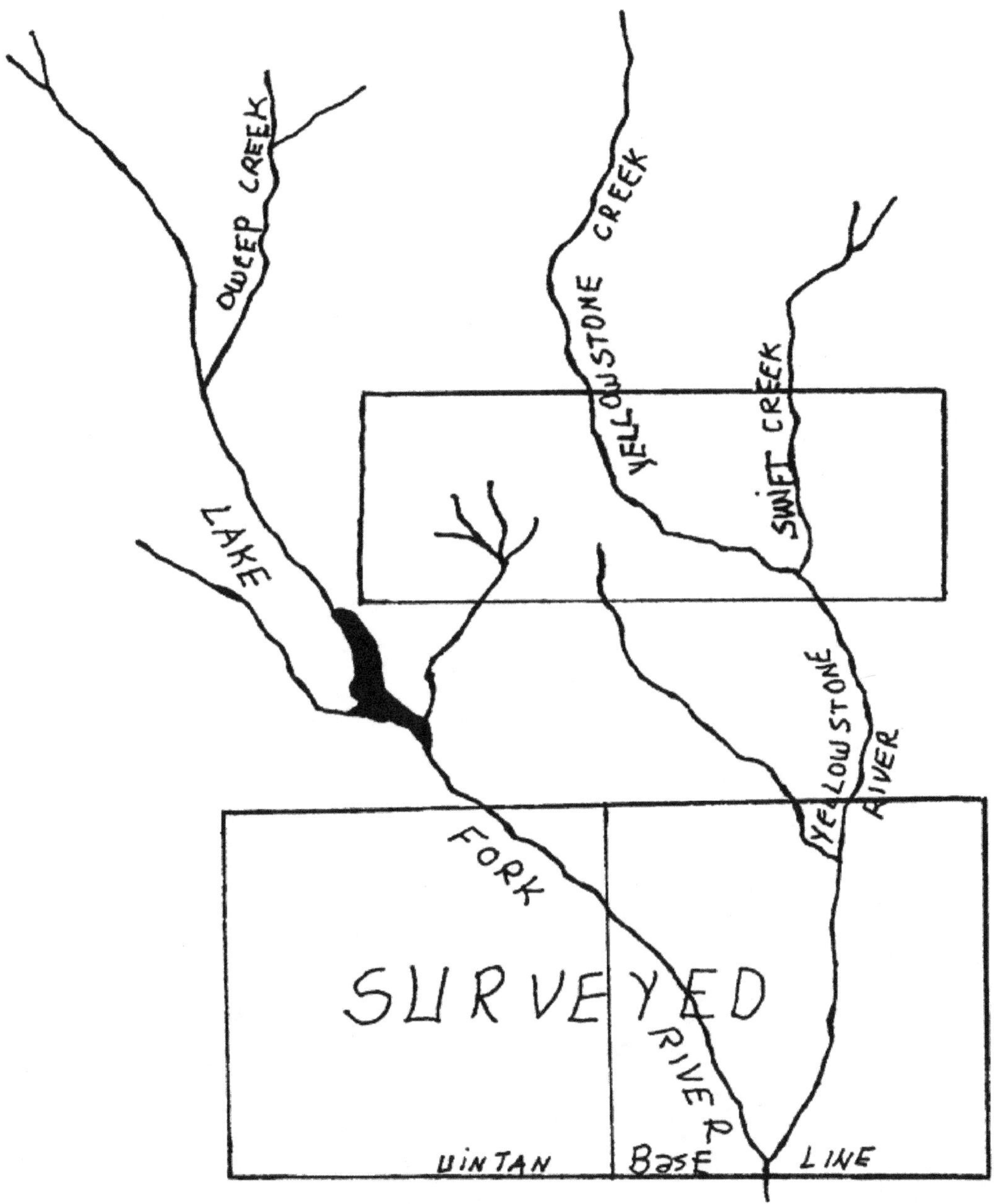

This is a map of the Moon Lake area as it was supposedly surveyed by Caleb Rhoades. This was one of Caleb's proposals to establish mining claims when the Native American land would have opened up to prospecting. He died just a few months short of seeing his dream fulfilled. The lower surveyed area would have encompassed the Twin Pots area and west, while the upper square would have taken in the mines on Swift Creek and the Yellowstone.

Rock Creek

There has probably been more written about Rock Creek than any other landmark in the mysterious Uintas. There are people alive today who can tell you fascinating experiences of their ancestors in this vast and lengthy canyon. So much has been said that it is difficult to decide just where to begin. Let's start at the southern end and work our way north. The area shown here is on Native American property.

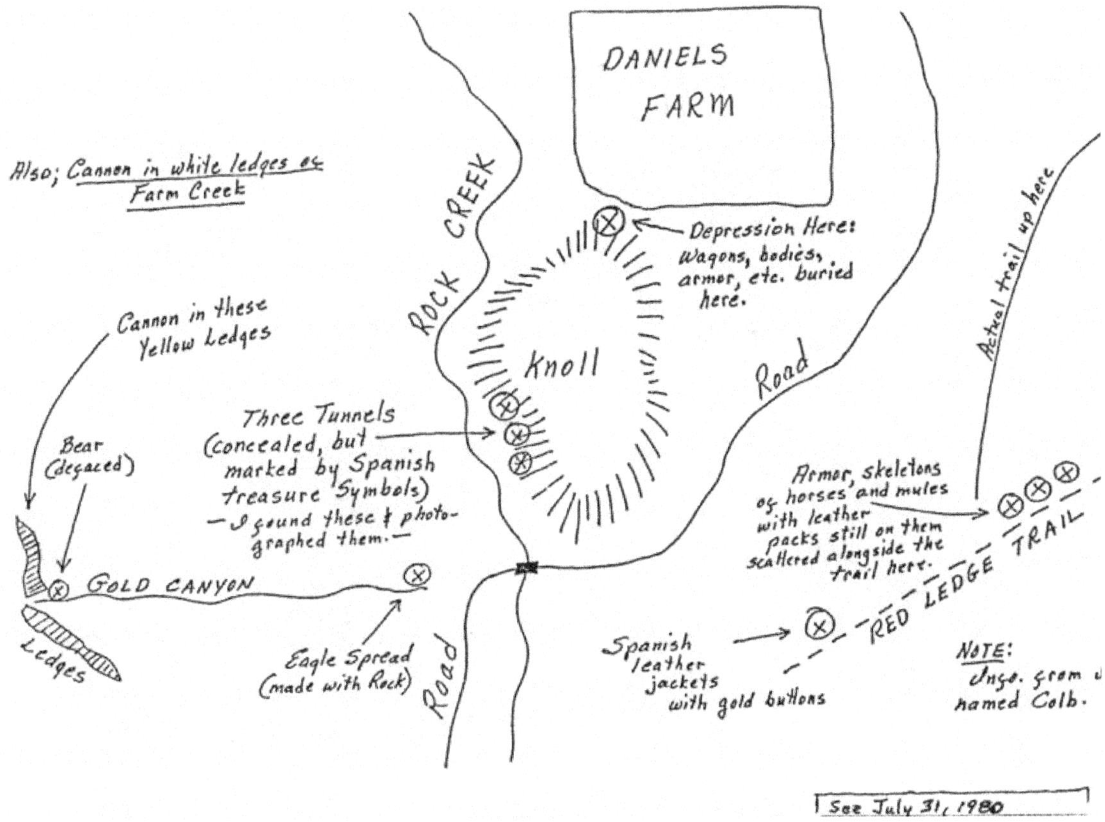

This map was drawn by Gale Rhoades, and I thank his daughter Paula for allowing me to include it and other maps and documents in this book. The area shown here is on Native American land.

A very large pine tree once stood at the base of Treasure Hill (marked "the knoll" on the map). Three deep crosses were easily seen in its thick bark. They represented two church treasures and one Jesuit treasure that were once concealed inside Treasure Hill beyond the water traps (see illustration on the next page).

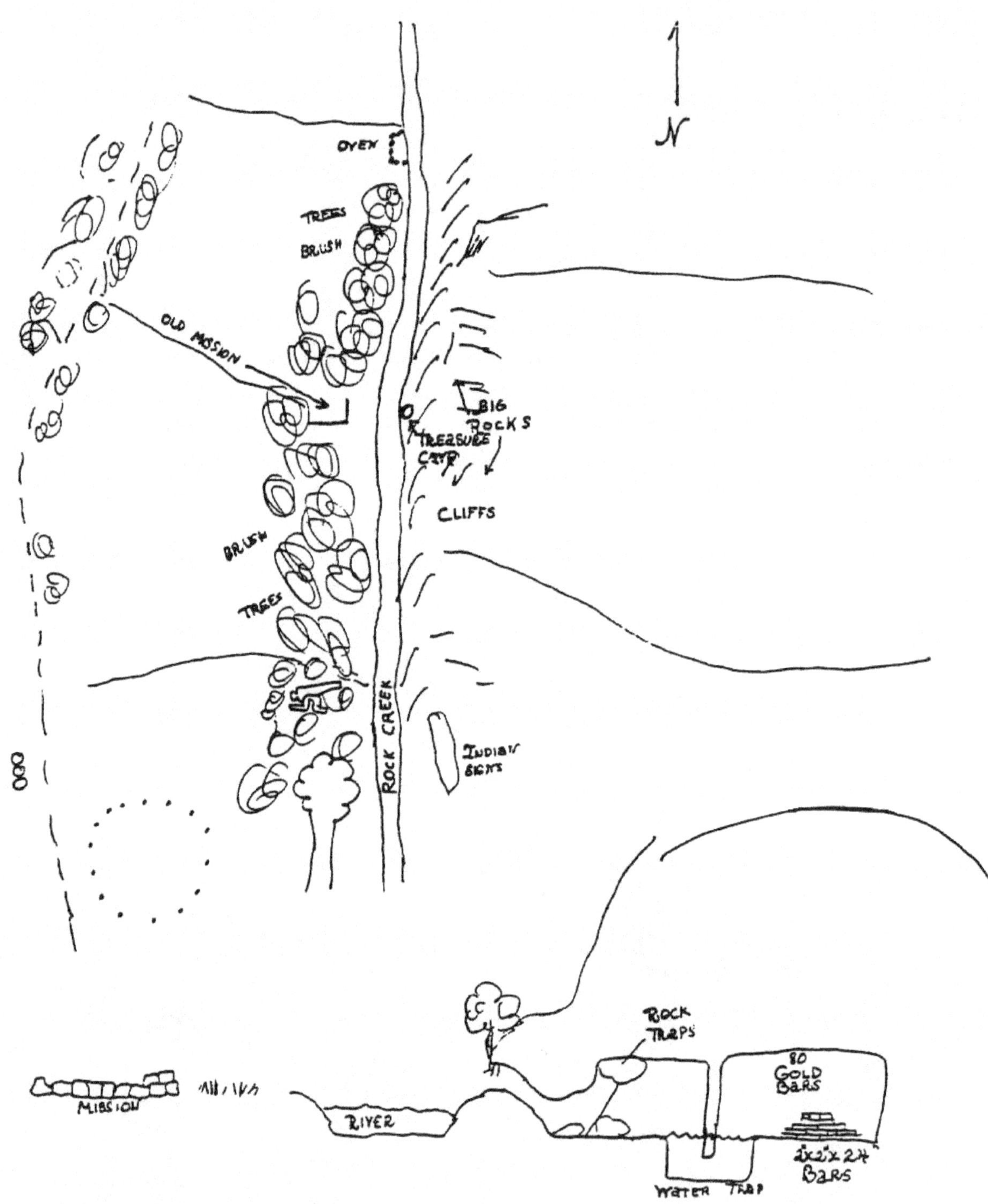

The treasure chambers in Treasure Hill could be entered via water traps. The Native Americans have removed the contents to keep the gold bars safe from looters. According to a close friend, there are many more like this one up and down Rock Creek. Some are on Native American property and some aren't.

Rock Creek is abundant with Spanish and Native American symbols like these sent to me by Patsy Sursa. Her daughter, Cori Wilde, is beside a cross symbol in the lower picture. Patsy says there are about fifty symbols in this area of Rock Creek.

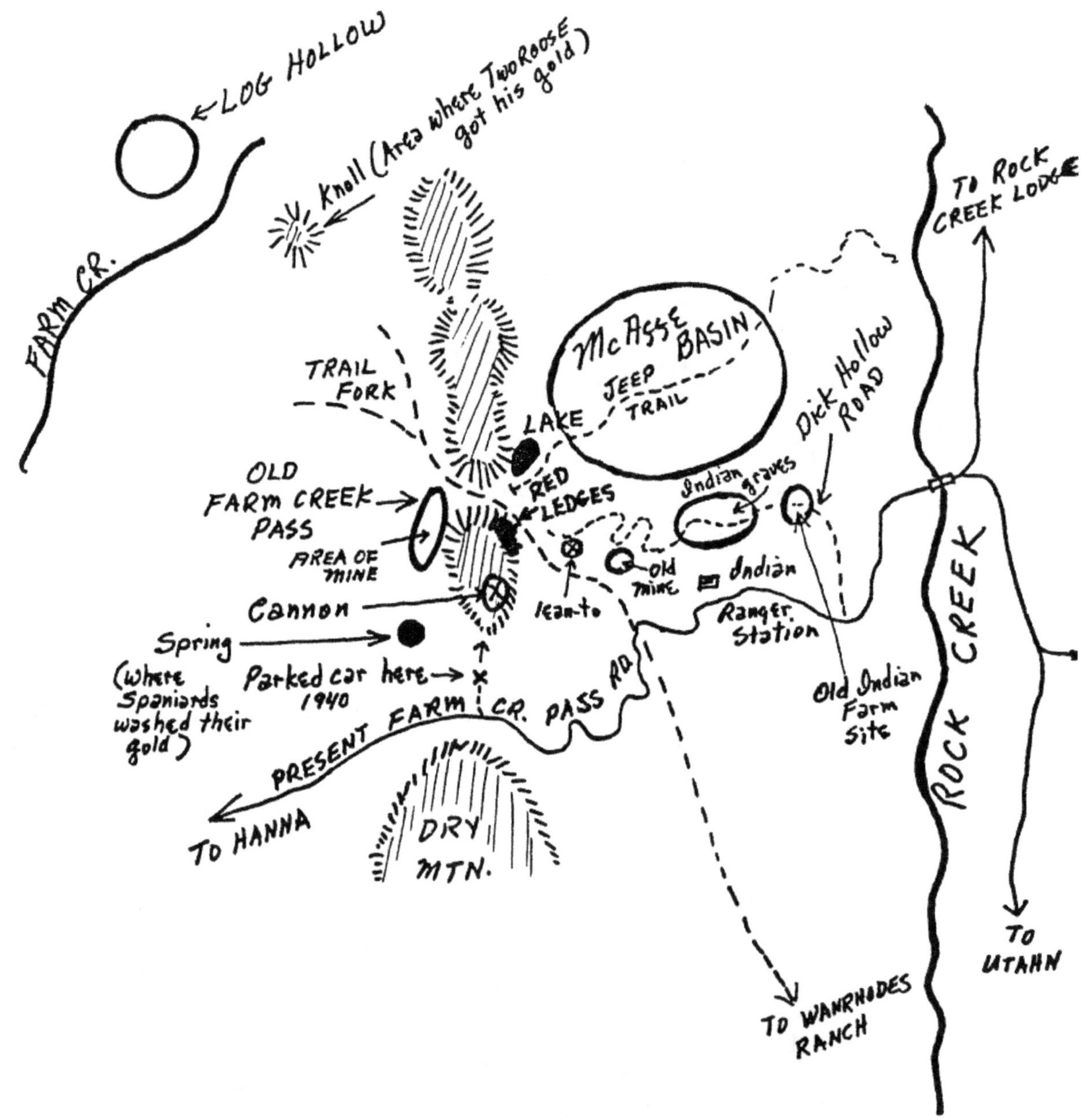

This is another one of Gale's maps. Ed called this one "Good." Gale's explanation follows.

Gale Rhoades Document

Author's Note

This has been rewritten because the original was in such poor shape, making it difficult to read. Other than the fact that I didn't capitalize each word, it is written word for word as the original. The copy of this information that Ed gave to me said "To Show Don Foot" at the top. Whether that means Ed was going to show Don or someone else, I don't know. Now for Gale's explanation.

This map shows other items of interest in the Farm Creek Pass area. Each item on the map is explained below.

1. Old trail, pass, cannon, and knolls were pointed out by old Chief Tworoose to Don Foote about 1920.
    a. The Trail (a rough wagon road) was in use during Caleb Rhoades's day. (The present Farm Creek road was not put in across the pass to the south until about the 1930s.)
    b. The Knolls was where young Don Foote held the reins of Tworoose's horse. Tworoose was gone only about 15 to 20 minutes and returned with two different pieces of gold. Tworoose had brushed his trail away behind him, but Foote is certain one of Rhoades' old mines lies here, somewhere near the knolls and at the base of a small ledge.

2. Old mine near "old" Farm Creek pass. This information was gleaned from several Indians including the Indians called Crazy Man and Brooker T. Washington.

3. Spring and parked car here: It was here that Mr. Larsen, his son, Edwin J. Larsen, and Happy Jack parked their car in 1940 and walked north searching small ravines to the east for Rhoades's mine. The spring is said to be the place where the Spaniards washed their gold. A gold nugget was found near its edge.

4. Dick Hollow information: Gleaned mainly from my own searches.

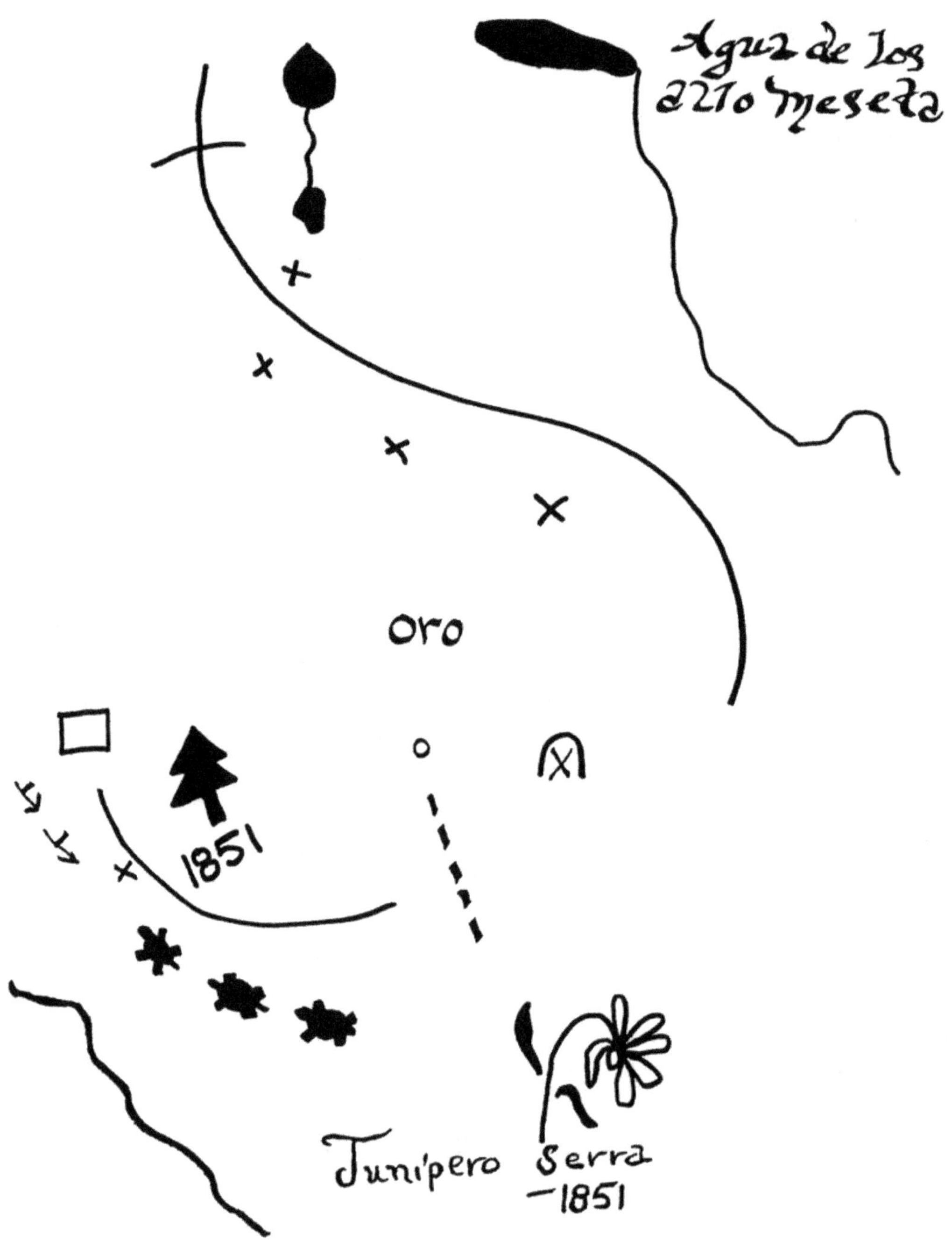

This map may be authentic. However, it may have been altered. Specific portions may have been reversed or flipped from right to left. There is a location in Rock Creek Canyon where arrows like these lead from the right or east and turtles lead from the left.

During a presentation given by Lisa Boren, Kerry Boren's wife, at BYU on May 11, 2001, Lisa stated, "Not all maps dealing with lost mines and buried treasure are genuine; in fact, it would be safe to say that most are not. Not only are many of them fraudulent, but many others are altered or otherwise coded to prevent theft. For example, when my husband and his cousin Gale Rhoades were writing *Footprints in the Wilderness,* they frequently altered maps by removing certain symbols or landmarks, reversing images, and so forth. Later, some of these maps were stolen and soon appeared in magazines and books as 'originals.'"

Here are pictures of the location in Rock Creek Canyon where these symbols exist as I have described them. Notice the unique arrows coming from the right and the turtles from the left. A friend of mine named Woody Olsen told me about them and asked me not to divulge the exact location of these symbols until we have fully checked them out.

Woody has consented to share these photos of this area in Rock Creek Canyon where symbols still abound. If we read them correctly there may even be a treasure nearby. Wouldn't that be nice? Thanks, Woody.

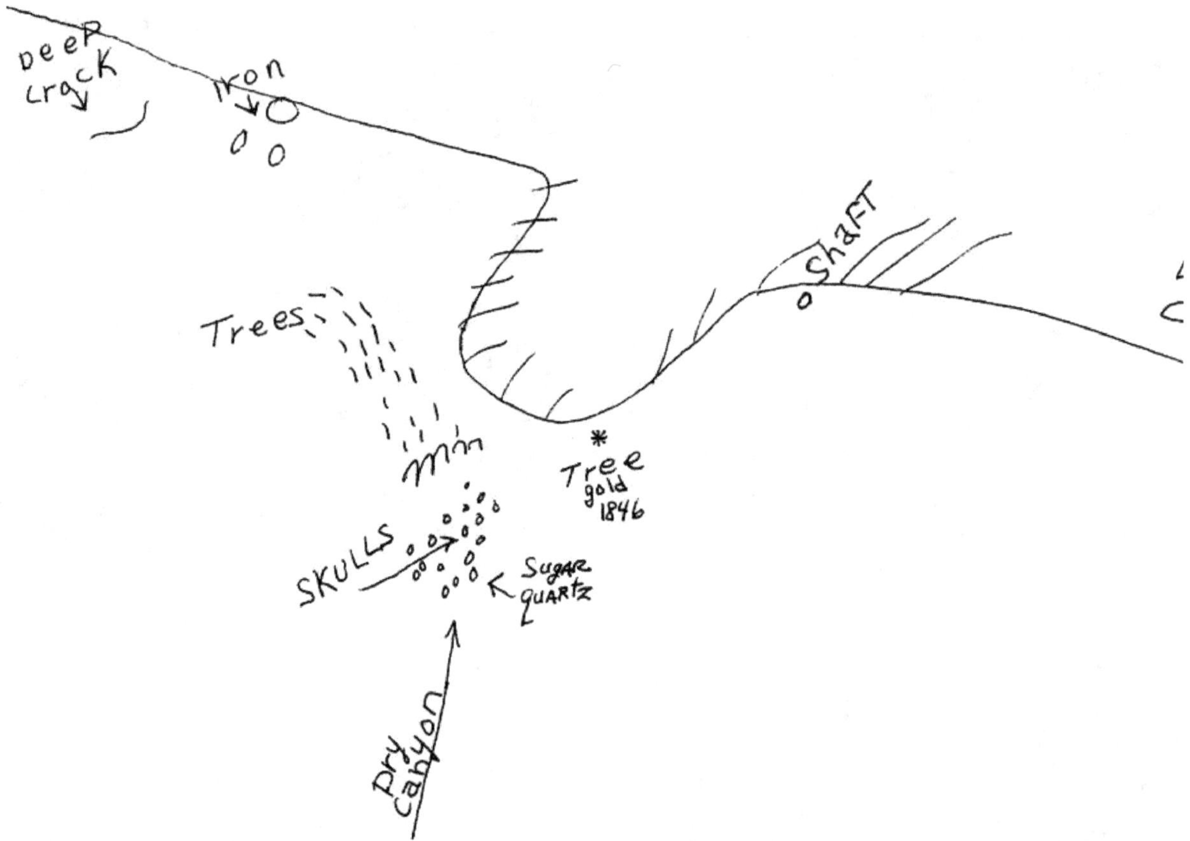

This map, as well as the next two, appears to deal with the massacre at the head of Dry Canyon. It has taken me a very long time, as it sometimes does, to put two and two together and come up with the correct answer. In other words, this is the same location Ed had told me about several times. It's the same area where he found a large stone ax head which may have been used during the massacre. He also said this was the place where Tworoose "slept," meaning he spent his last days on earth here. Tworoose wrote his life history here on rolls of paper and left it in bottles that Don Foote later recovered via helicopter. These rolls of paper were later buried with Don. Ed asked Don if he could take the history and have it translated into the English language since it was in Tworoose's native tongue, but Don wouldn't hear of it. He did, however, allow Ed to take the Killum Map and copy it. Tworoose made the Killum Map by placing small spots of his own blood on paper. The spots corresponded to mine locations, all in reference to an Aaron Daniel's mine in the middle of the rest of the mines. Because I understood where the map was drawn, I believed it pertained to the Kidney Basin region just north of the place where Tworoose died.

It is generally believed that the Killum Map refers to an area near the head of Blind Stream. I was taken to a spot in Blind Stream Canyon where I was shown very old pieces of someone's shoe and other items. I was told by my Native American guide that this was a massacre site also (in Blind Stream), and he told me other information which he asked me not to divulge. Along with others, I now believe the Killum Map pertains to Blind Stream because of the information he shared with me. The question still remains in my mind as to why Tworoose drew a map of Blind Steam from the head of Dry Canyon. I suppose he must have had the Killum area memorized and didn't need to be in Blind Stream Canyon to draw a map of it. (To view the Killum Map, see page 164).

This map shows Moon Lake to the east and the drainage into it. It also shows the head of Dry Canyon and the associated mines.

The asterisk that I have placed on this map, the previous map, and the next map mark what appears to be one and the same location—the famed Pine Mine, of which Benjamin Hart Bullock was accused and then exonerated of destroying with a Caterpillar tractor. (See *Footprints in the Wilderness, A History of the Lost Rhoades Mines.*, page 396.) One map says 1856 while the other says 1846. They appear to be the same mine even with the discrepancy. There is now an open pit quarry there. If there was ever a shaft it must be covered because we couldn't find it when Dan Howard and I visited the site in August of 2003.

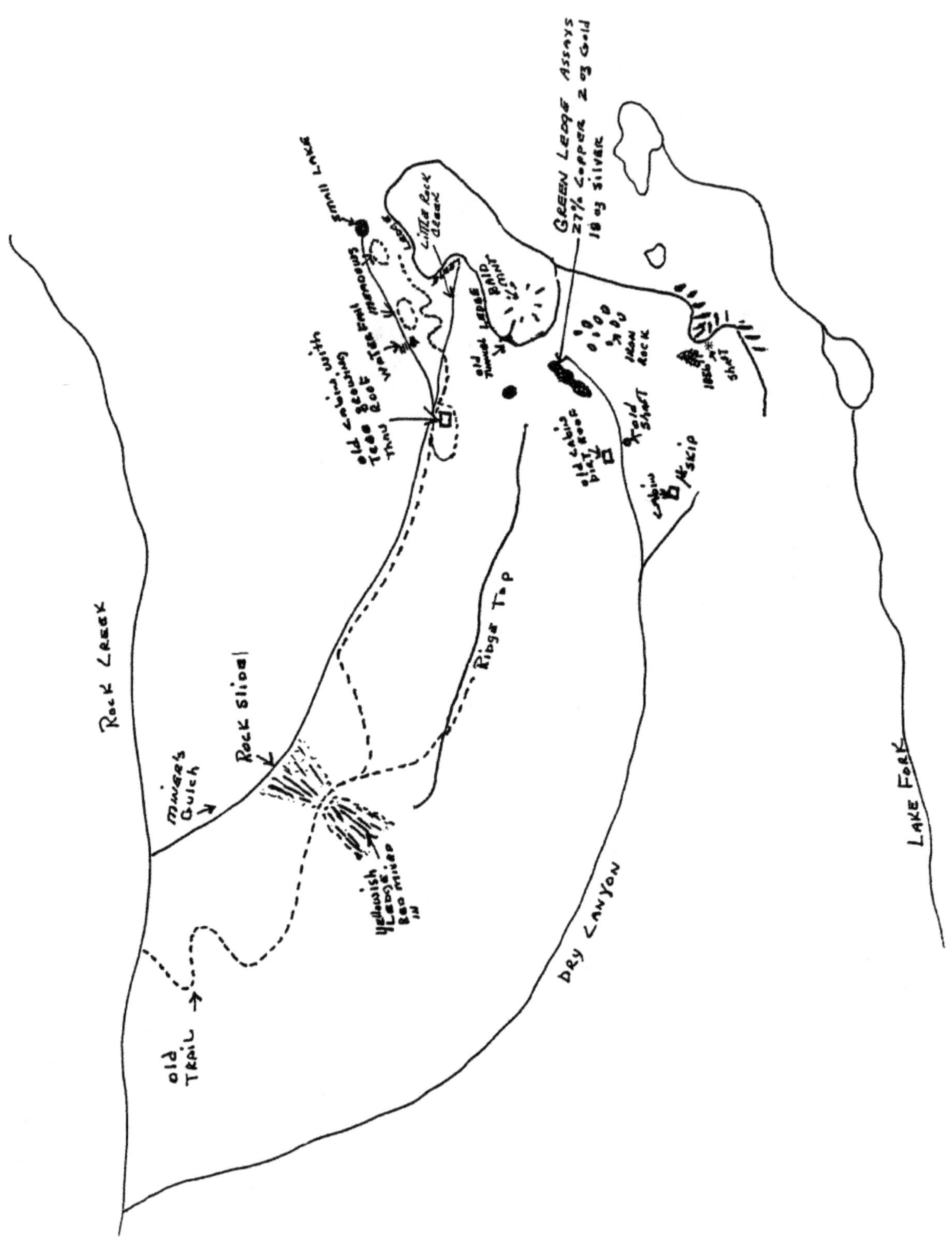
Rock Creek
Rock Slide
Miners Gulch
old Trail
Ridge Top
Dry Canyon
Lake Fork
Small Lake
Green Ledge Assays
27% Copper 2 oz Gold
18 oz Silver
Iron Rock
old Shaft
Skip
Yellowish Ledge
Red mixed in

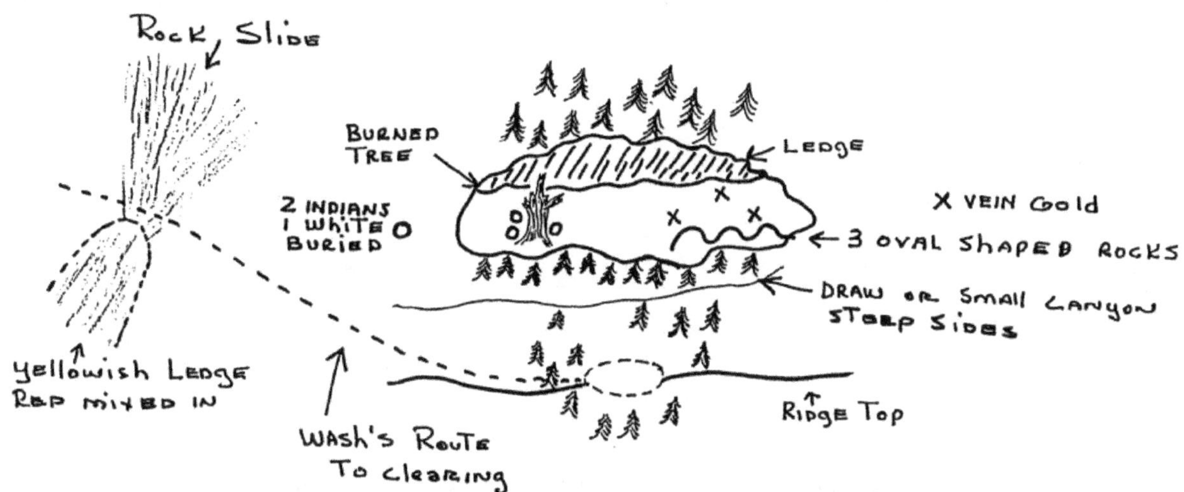

This map and the one on the facing page go hand in hand. This one indicates the route taken by a Native American named Henry Wash. The oval-shaped rocks are mentioned in the Thomas letter that follows as "three rock(s) like bad snake back." Thomas was actually Waubin Wanzites.

Gale Rhoades Document

Author's Note

This has been rewritten because the original was in such poor shape and very difficult to read. Other than the fact that I didn't capitalize each word, it is written word for word as the original. Now for Gale's information.

Document:
Letter from an Indian

Date:
December 30, 1969

My friend,

To answer you letter Yopana told my father you savvy where other side big sattle. Me to be full blooded Ute Indian. I no want to talk about my name—about 1912 summers we goes to Sattle Farm Creek road lots talk about money rock—to south carved bears both sides mtn. Sacred mine Red Man gold.

They say water run under for bridge nature on trail by black burn rock (coal). Maybe so mepooch [little] canyon then 1 more north sides, place look like horse foot track some red paint mud and mepooch blue–mepooch ledge or rock. In horses track horses track look like correll gold in there spring closet. Old Indian camp in quake trees lots grass-rock brown gold look like honey comb. Rock creek where spoint mepooch iron doors. Maybe gold bars—up rocky creek east side three rock like bad snake back. Dinosor—gold north east under ledge- big iron doors- summer time I come show don't to tell my people or dead me. Rhoades mine think west over pass, I see one pile rocks cover—large Spanish guns. Up dry rocks cover with rock—I think no more. See you summer time.

A friend
Thomas (aka Waubin Q. Wanzites)

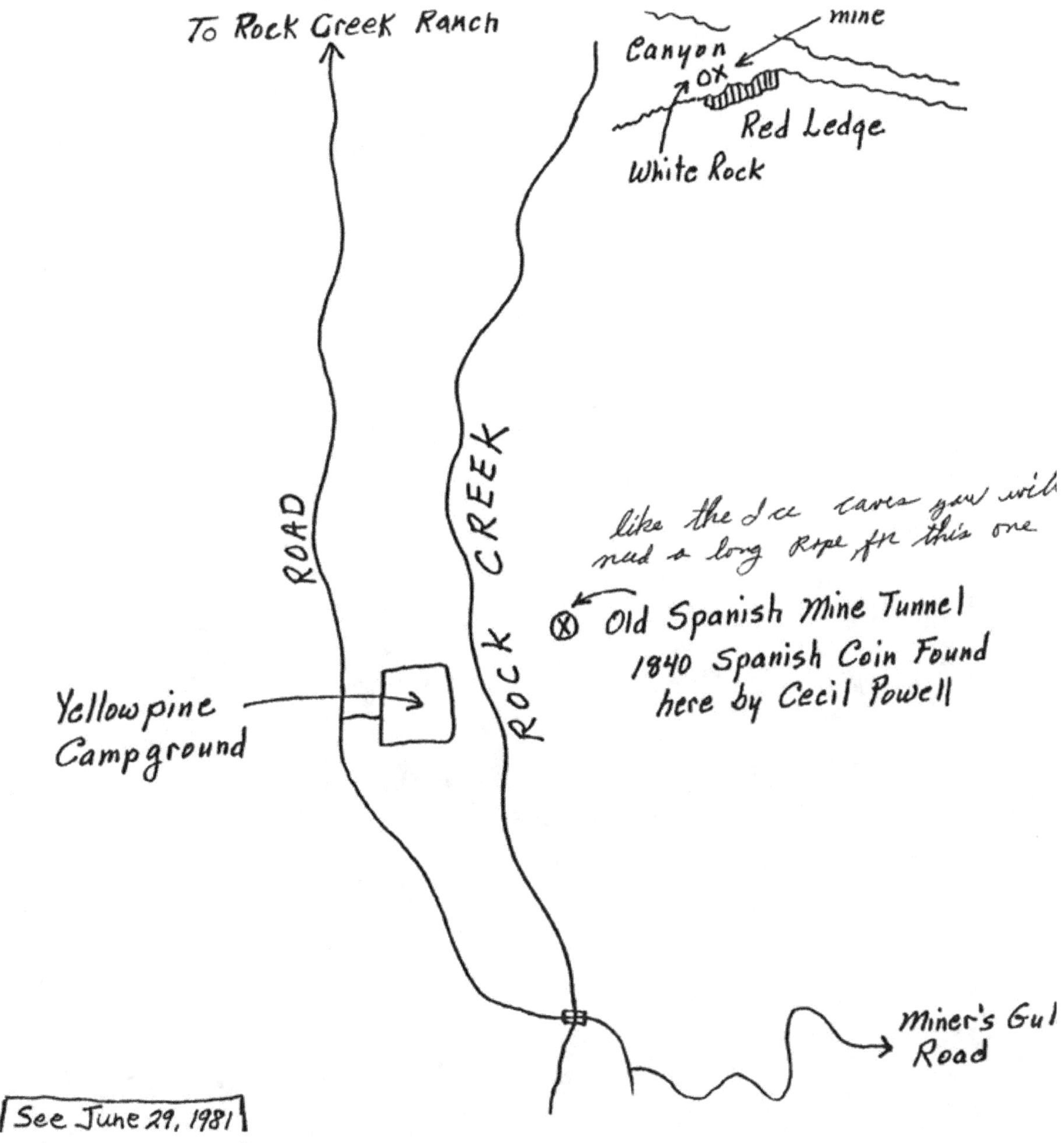

This map depicts the area just below the upper Stillwater dam. Gale drew this map, and Ed wrote the lines about a long rope. I received this copy from Ed.

In the late 1980s, I checked on the mine near the spot where an 1840 Spanish coin was found (shown on this map) while I was looking for Richard Park of Orem, Utah, and his silver mines. The mine where the coin was found was just a depression in the ground when I saw it. Much effort would be required to reopen it.

This map is shown in *Lost Gold of the Uintah, The Rest of the Story*. Gale calls it Waubin Wanzite's map, but I don't know why. I assume the information about the mine shown on the map came from Waubin. The next two maps record information I have collected over the years.

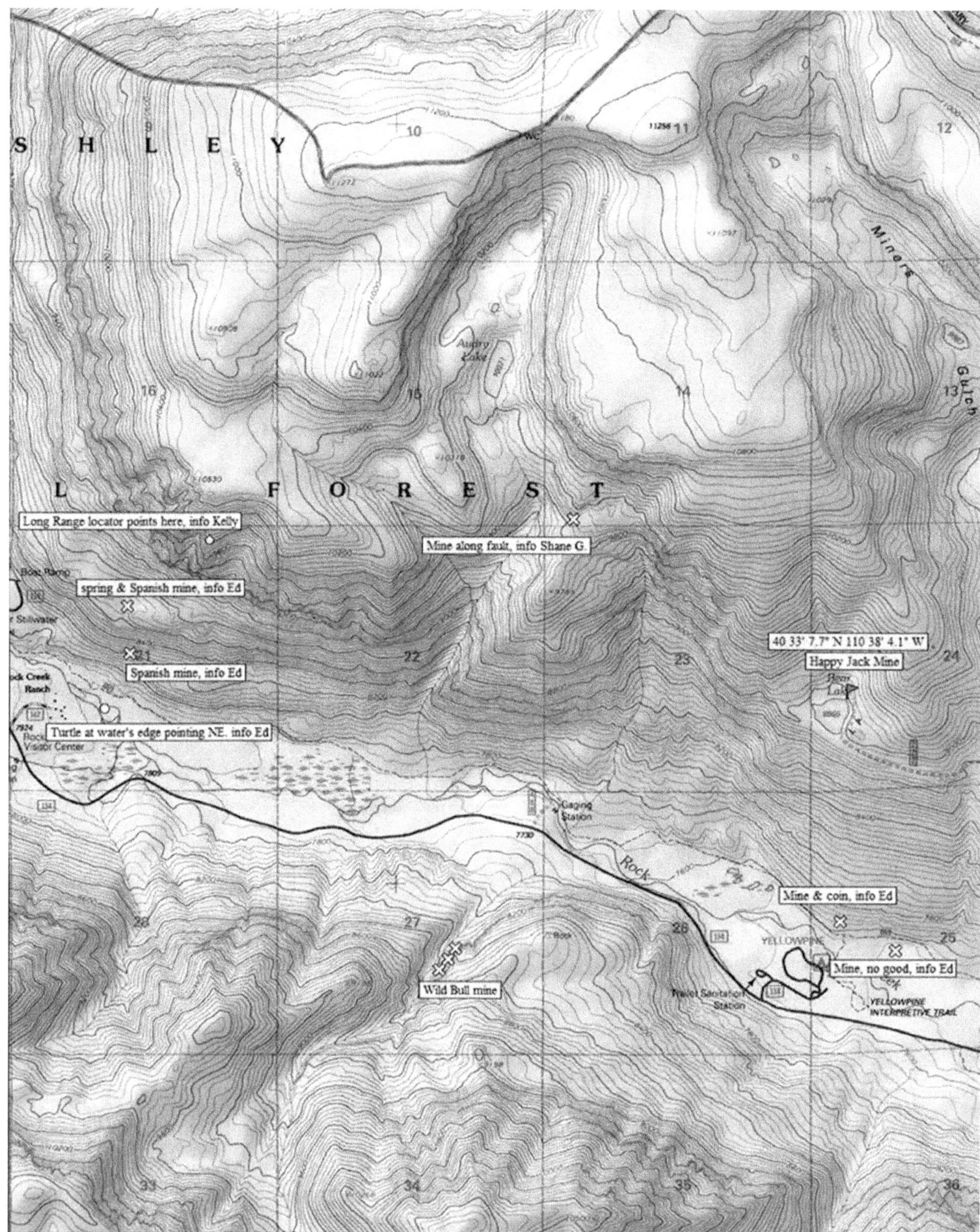

This map is of the Yellow Pine Camp Ground and upper Stillwater area of Rock Creek Canyon. It's a graphic record of information I have collected over the years from various sources and from my own hikes. Map created with TOPO! © National Geographic Maps.

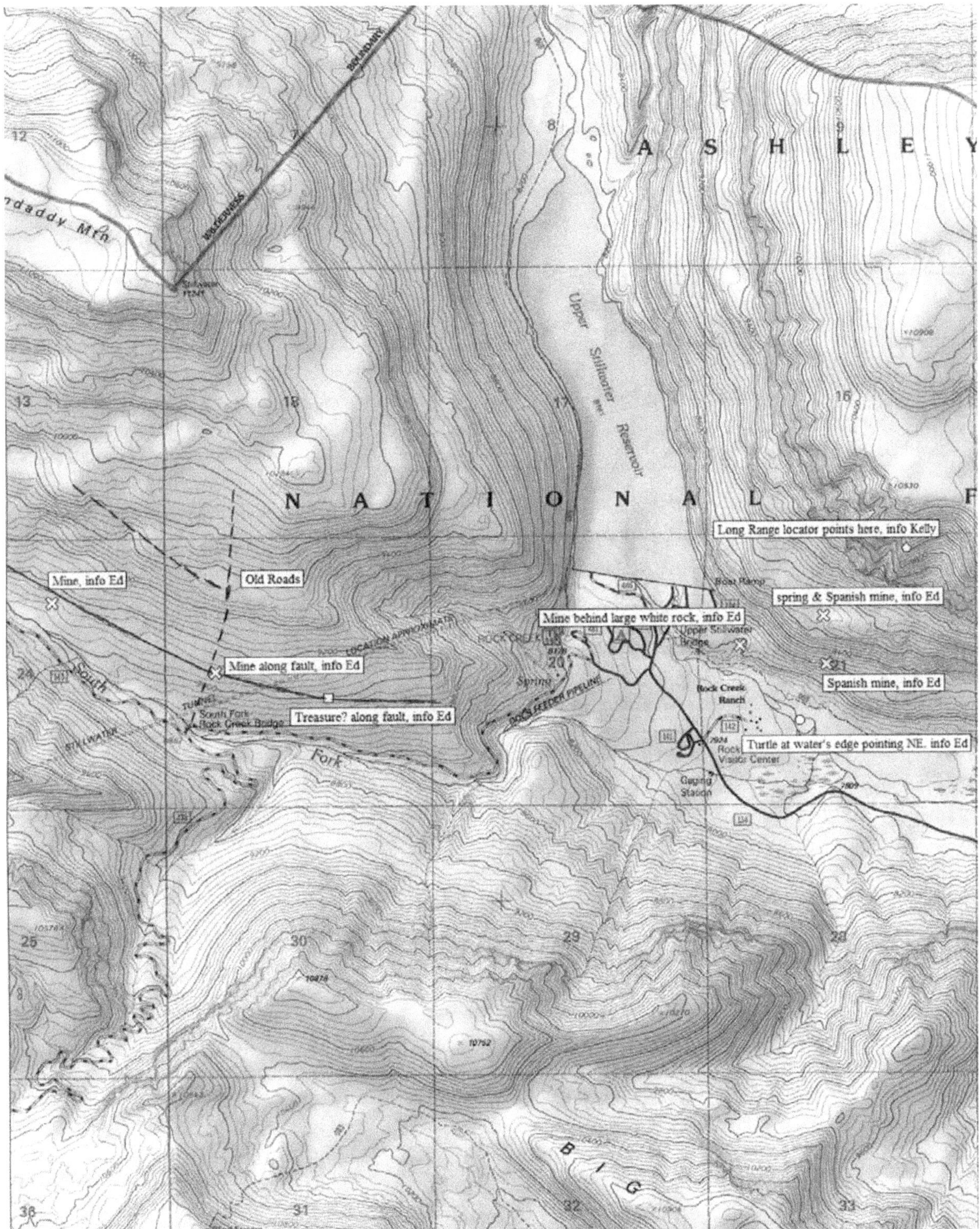

This map contains information about the upper Stillwater area of Rock Creek Canyon. There are several items of interest on this map that I haven't had time to verify. Map created with TOPO! © National Geographic Maps.

Gale Rhoades spent countless hours exploring Rock Creek. This is documentation of a few of his discoveries. His explanation follows on the next page.

Gale Rhoades Document

Author's Note

This has been rewritten because the original was in such poor shape and very difficult to read. Other than the fact that I didn't capitalize each word, it is written word for word as the original. Now for Gale's explanation.

This map merely shows general information gleaned from reliable sources and from my own searches during 1961–1964.

The numbers on this map mean:

1. Bad water: We became sick and broke out in a rash upon drinking and washing in this spring-like stream.
2. Large cave: This cave appears, from above the switchbacks, to be no more than a large pocket in the side of the ledges, but a few feet in, it turns to the right and continues into the depths of the mountain. No one that I know of has ever went more than ¼ mile into this cave and at least one man has nearly lost his life in it.
3. Small hole: This hole, near a small ridge of protruding rock, can only be seen from one spot on the switchbacks and even then one must look closely to see it. We wanted to check it out but could not locate it by dropping off the top of the mountain and my cousin, Garry Rhodes, nearly fell from the ledges to his death during our one and only attempt.
4. Natural tunnel through ledge: This tunnel appears to be made by nature. However, since a 2 to 3 foot vein of white quartz or calsite (calcite) runs vertically very near this tunnel, I have always wanted to go up with ropes and drop into the tunnel from the upper side (one can not climb into the tunnel from the lower side) to investigate whether or not a man-made tunnel runs off the natural tunnel and into the mountain ledge. I've never done this, but it should be checked out.
5. Black shale: This shale contains traces of gold.
6. Old mine: Old shaft found here in poor shape.
7. Cave: A note was found in this cave supposedly written by C. B. Rhoades.

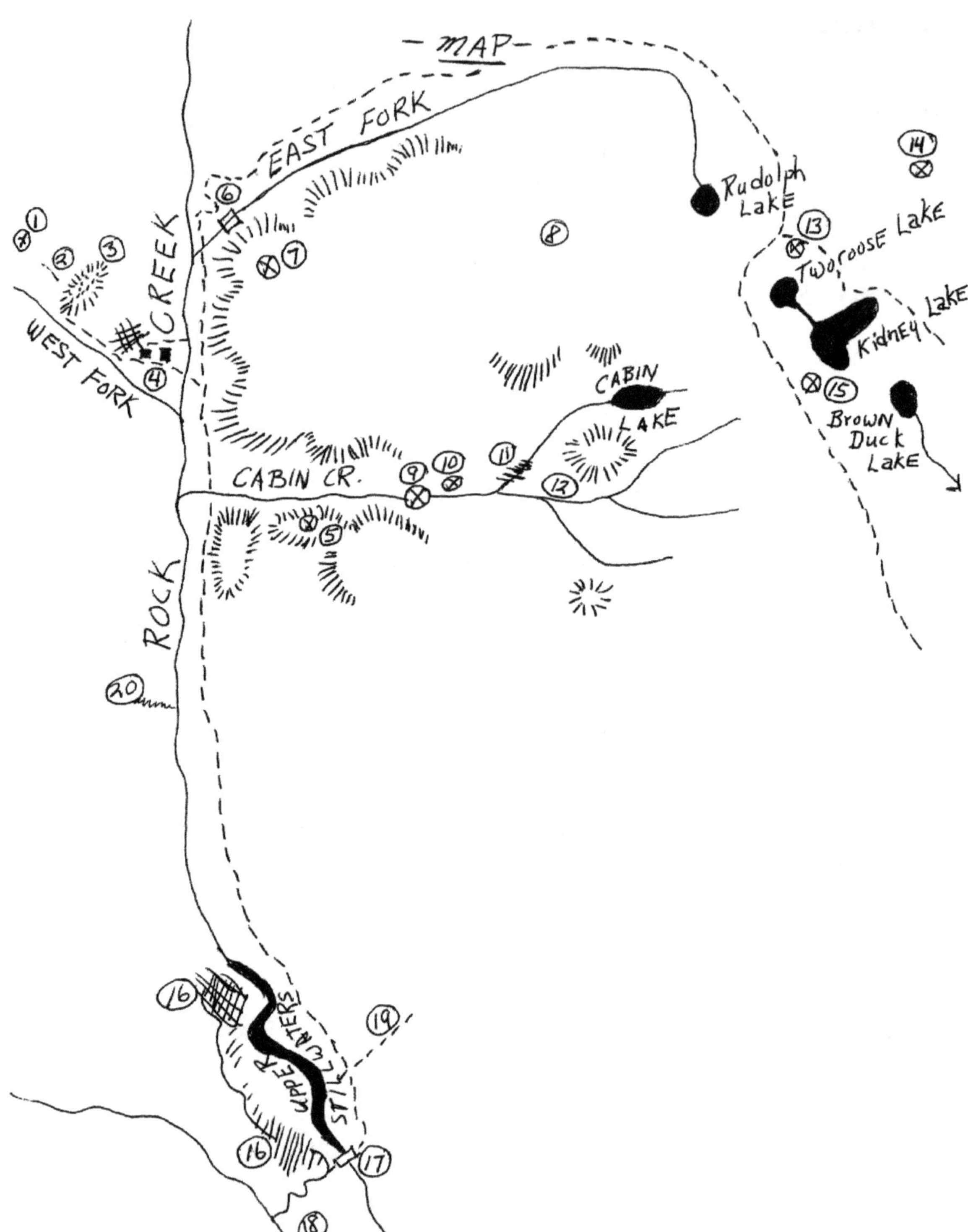

Gale's explanation for this map follows on the next page.

Gale Rhoades Document

Author's Note

This has been rewritten because the original was in such poor shape and very difficult to read. Other than the fact that I didn't capitalize each word, it is written word for word as the original. Now for Gale's explanation.

This map is of the Rock Creek-Cabin Creek region and in this I've attempted to show special points of interest known and discovered by myself from 1963 to 1975. The explanation of the numbers on the map are as follows:

1. Silver claims–Located by Leland Stevenson of Altamont, Utah.
2. Fault–Running southeast.
3. Ridge–With silver ore showing through.
4. Two old cabins near small clearing–One has nearly rotted away, the other has been burned down.
5. Silver vein–Once located by Clark Powell, Jr. and myself.
6. Area of iron door mine–Located by a fisherman (A doctor from back east by the name of Rhodes) in 1925. Visited by Happy Young, who was run away from it by Native Americans after they cut three of his fingers off. Very near waterfall on East Fork.
7. Old mine–Seen through binoculars from an adjacent mountain by Ron Bennion in 1966.
8. Very old trail–Photo of trail taken in 1963 as we flew over the region. Wide enough for wagons. Probably built by Spaniards.
9. Home-made gold pan–First located by June Balaich and later found by Clark Powell, Jr. and myself. I placed it upon a rock near the creek. I suppose it is still there.
10. Timber wolf–Spotted at close range by my brother, Ted, and myself in 1967.
11. Slide rock.
12. Cut in knoll–Square shaped, about the size of an old cabin floor. No visible evidence of cabin remains, but there certainly must have been one there many years ago. This site should be checked out with a metal detector, as the Spaniards very often buried their gold and silver bars in the dirt floors of their cabins.
13. Tree–With Spanish symbols.
14. Area of old mine shaft found by a friend of E. J. Larsen.
15. Area of old mine tunnel–Found and lost again by two Uinta Basin cattlemen.
16. Upper Stillwater campgrounds.
17. Footbridge.
18. Rock creek lodge.
19. Fence.
20. Waterfall.

This photograph was taken July 24, 2002. The view is from the head of Blind Stream looking north to northeast. Squaw Basin can be seen in the distance near the center of the picture. The upper reaches of Cabin Creek are near the right side of this photo and if you look closely you can see the distinct top of a hill shaped like a pyramid.

Verl Iorg, whom I was with when this picture was taken, told me we were looking at a pyramid, but I thought he might be pulling my leg. Until then, I had never heard of such things as pyramids in the Uinta Mountains. As far as I knew, they only existed in places like Egypt and South and Central America. Since this time of awakening, I have accumulated more information on the subject. An entire chapter, and then some, could be written about the ancient history of Utah and a possible link to the ancient Mayan people. I have very little proof that there is a connection, so I will forego any opportunity to elaborate at this time.

It is my goal to visit this mysterious hill, which could possibly be a pyramid, and validate our findings. When I visit it, I will attempt to determine whether it is a natural mound or if it has been altered to appear like a pyramid. Of course I will also look for openings while I'm there, and document my findings.

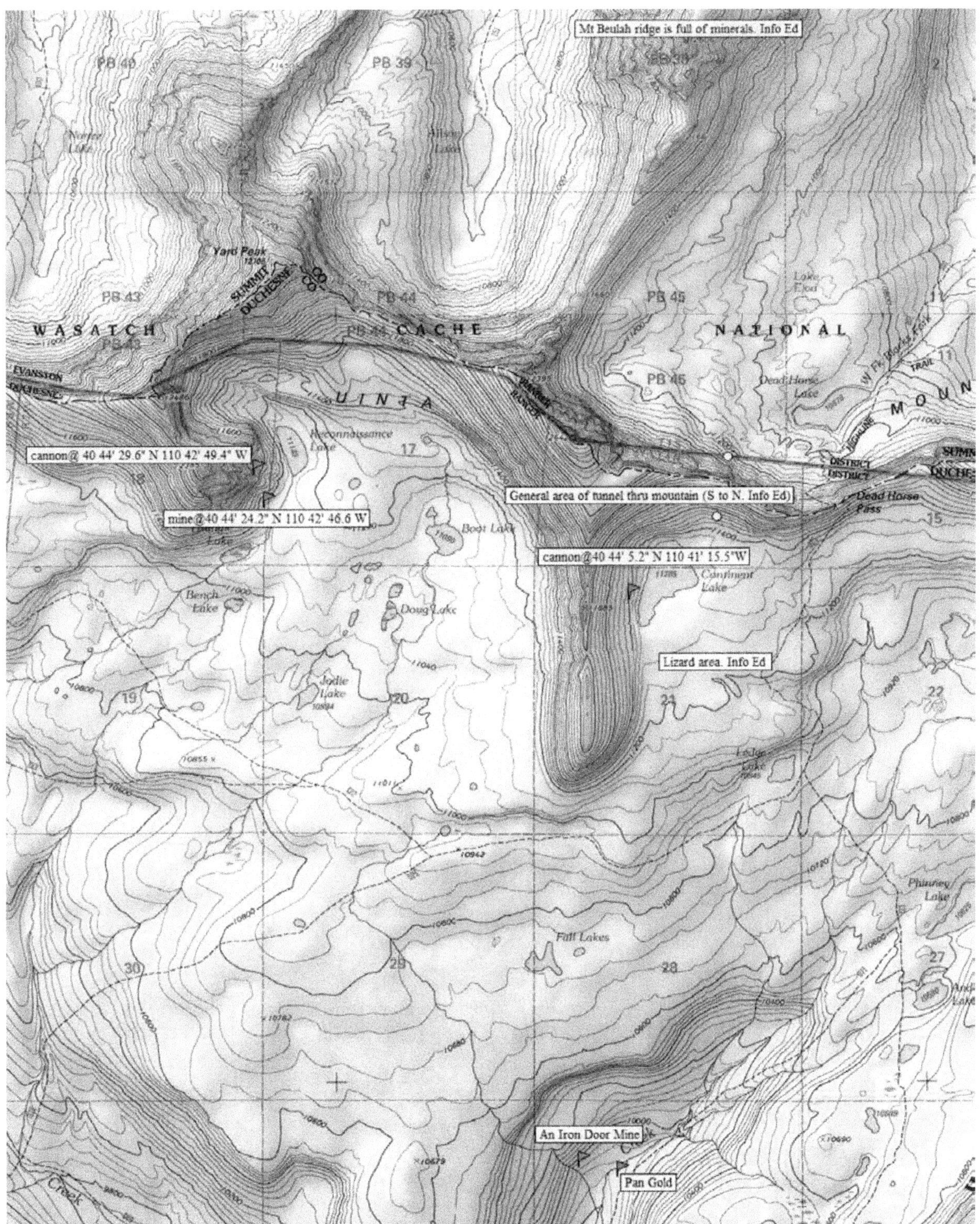

This map displays the upper reaches of Fall Creek and Rock Creek. Most of the information marked on it came from Ed with the balance from Randy Lewis. Both are reliable sources. Map created with TOPO! © National Geographic Maps.

Blind Stream

Symbols and mines abound in Blind Stream Canyon. As you leave Hanna and travel through Native American land onto Forest Service land, you find yourself in the midst of a vast array of history. There are so many markers and symbols that it almost makes your head spin.

It has been enjoyable to see and interpret the signs as much as I have. I have found that stacked rows of rocks as well as cut lines on trees can mark veins or other important things. It's important to study and ponder the meanings if we are going to put the puzzle together. As I learn how to interpret these symbols and markers, it's exciting to have so much at my fingertips in a location like Blind Stream. It's almost like a knowledge-hungry student in a vast library of books, or a sugar-starved kid in a candy store.

We have access to much knowledge and history in this, and many other locations, in the vast and inviting Uinta Mountains. Blind Stream is surely one of my favorite locations for the study of our Spanish history.

This is one of two bird symbols I found near the Blind Stream road. There are several cat face symbols as well as other symbols in this intriguing canyon.

Sometimes the markers are easy to find and interpret. Other times it becomes a challenge. The line of rocks on the ground and the line carved in the tree in the upper photos both appear to be marking a vein.

The arrow beside this cat face led me around the hill, following a manmade ditch, to a gully where I believe a mine and smelter once existed.

I found cat faces here as I have in many locations in the Uinta Mountains and the Uinta Basin. This time it was very easy to locate the caved-in mine a short distance past the symbol. This reaffirmed my hunch that the cat faces are markers for the mines.

In this picture, I'm standing next to a mine dump that I saw immediately when I looked past the cat face symbol.

Here are two trees that were used as markers. One has wooden spikes sticking out of it, and the other has a metal file driven into it. Markers like these often indicate the direction of travel to reach a mine, a vein, or they could even be on top of a vein. In this case, they led to several mines dug along a nearby vein. I have found spikes driven into trees in other areas too, such as Lightning Ridge and Mosby Mountain.

This is one of several mines along that vein that I found by following the spikes.

Yes, there certainly is a lot to see and find in Blind Stream Canyon.

In the photo section of the book *Lost Gold of the Uintah, The Rest of the Story*, Gale says this thunderbird symbol was "found on the high ledges near the mouth of the south fork of Rock Creek." Since I read his description, I have looked for it whenever I was close to the south fork. See page 192 for my experience finding this thunderbird.

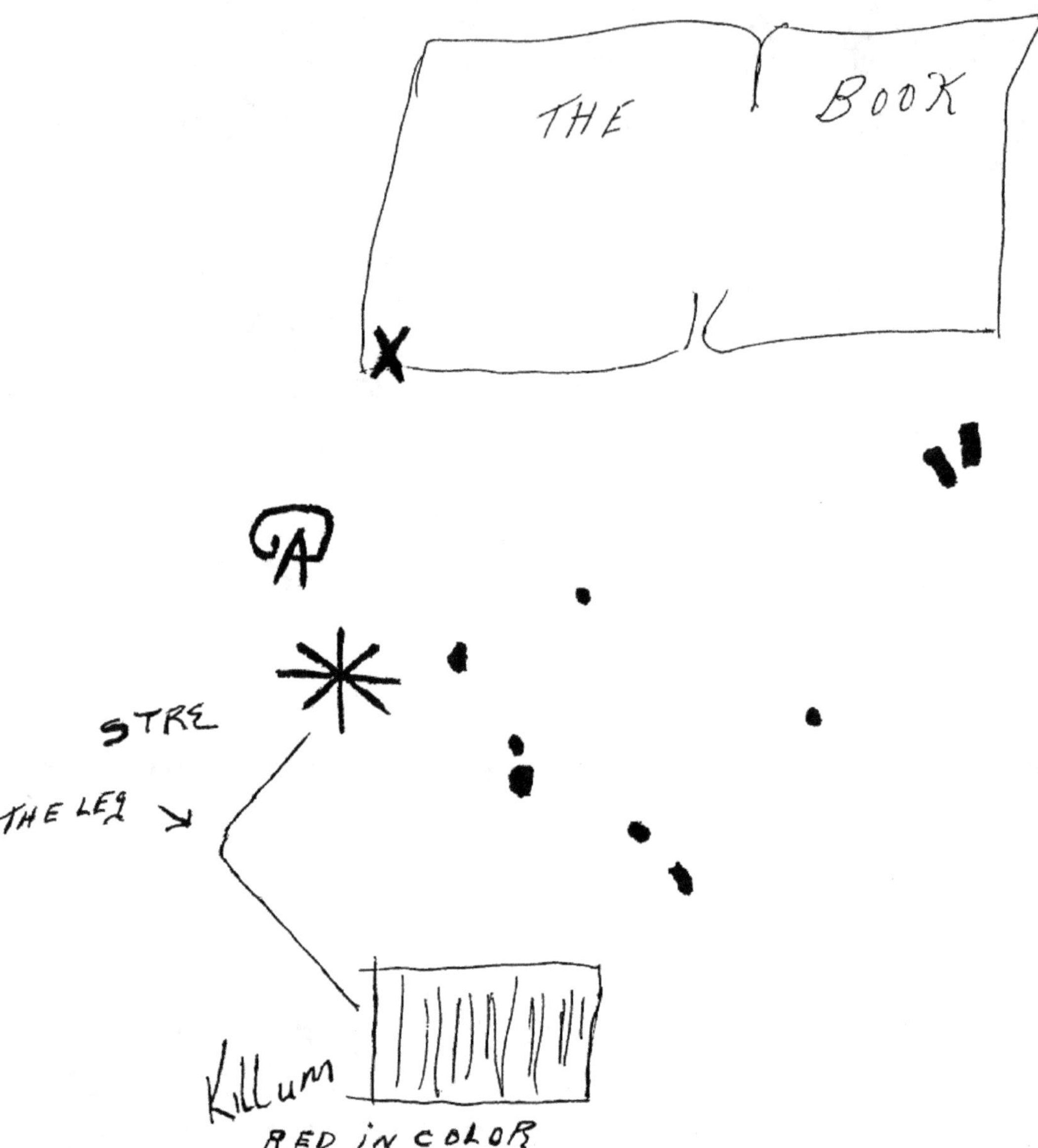

This is referred to as the Killum Map. Ed told me the dots indicate mine locations, and were made by Chief Tworoose blotting his blood onto a paper. For a long time, I believed it indicated Brown Duck Basin because that is near the area where Tworoose died, and I understand that's where he was when he made the map. Later I was shown an area in the upper reaches of Blind Stream Canyon with matching characteristics that led me to believe, as many others already did, that the map fits Blind Stream. The Killum Map shows a star. Ed told me that if I found the eagle I should look on top of that hill for a star. The puzzle pieces fit very well. (See page 144 for more details about the Killum Map.) I was asked by a Native American friend not to explain more than this, so I will let it rest.

Lightning Ridge and the Golden Staircase Mine

Lightning Ridge is synonymous with gold. Marked trees are easily found, as well as caves and even mines. Occasionally gold is discovered here either in wire form or in smaller amounts like the golden yellow of lightly buttered bread.

Much is hidden here just as it is in most locations. Occasionally, someone stumbles upon something life-changing. Such is the story of Binggeli and the Golden Staircase Mine.

The year was 1931, and Wilford Binggeli and Bill Nunley found themselves employed by the Forest Service, marking and cutting beetle-infested timber to keep the infestation at bay. They would never forget what occurred one particular fall day. As we retrace their footsteps, it wasn't difficult to find where they started, but past that the events are a blur.

Marked trees are abundant in Lightning Ridge. They serve as reminders of the area's history of gold mining.

It is agreed that their camp was at Cold Spring. It was late in the year and the weather was poor to start with that day and got steadily worse. Eventually it began to snow.

There were thirty men on the crew, and they were stretched out at one hundred foot intervals as they swept around the hillside on foot, examining each tree for problems. They kept their spacing as they gradually worked their way around the hill, traversing in and out of canyons and over ridges. But as often happens, the spacing changed and Binggeli and Nunley found themselves separated from the others and close to each other. As the temperature dropped, the snowfall increased. They decided it would be wise to return to camp. Binggeli led the way on their return hike. Visibility was poor and the snow was deepening and stinging as the wind whipped it into their faces. All of a sudden, the limbs, pine needles, and brush on the ground gave way as Binggeli fell into an unforeseen hole. Nunley found him shaken from the surprise but otherwise unhurt from the short fall.

As they looked around, they could tell Binggeli was standing on an incline of steps with vertical walls surrounding him. While looking into the hole and the darkness beyond, their eyes caught the yellow glint of something on the wall beside them. They examined it closely. There were two chocolate brown veins with unmistakable gold within, but that wasn't anything compared to the five-inch vein of white quartz between those chocolate veins and a solid gold stringer three-fourths of an inch wide that was sandwiched within the white quartz. The gold was breathtaking. They broke off a few pieces using whatever they could find as a tool. Then they quickly covered the hole to conceal it. Little did they realize that this would add to their future frustrations.

They made their way back to camp and found the crew loading up all of their belongings and preparing to leave for the season. It hadn't been planned that way, but the weather didn't permit many options.

Life was busy and Binggeli and Nunley didn't return to Lightning Ridge for a considerable length of time. By the time they did, their memories started playing tricks on them and, try as they might, they couldn't relocate the Golden Staircase anywhere except in their dreams.

It still lies right where they found it, with a few small logs and bough branches for a covering, just waiting for a lucky hiker to stumble upon it or maybe into it.

Ed wrote information on my map about the large Native American uprising against the Spanish and how many lives were lost. He told me these were Navahos. Possibly, there were other areas where the dead were buried because during the summer of 2005, I found what appeared to be a grave in the meadow a short distance from where Ed indicates the battle occurred. He also marked the location where he believed the Golden Staircase Mine was.

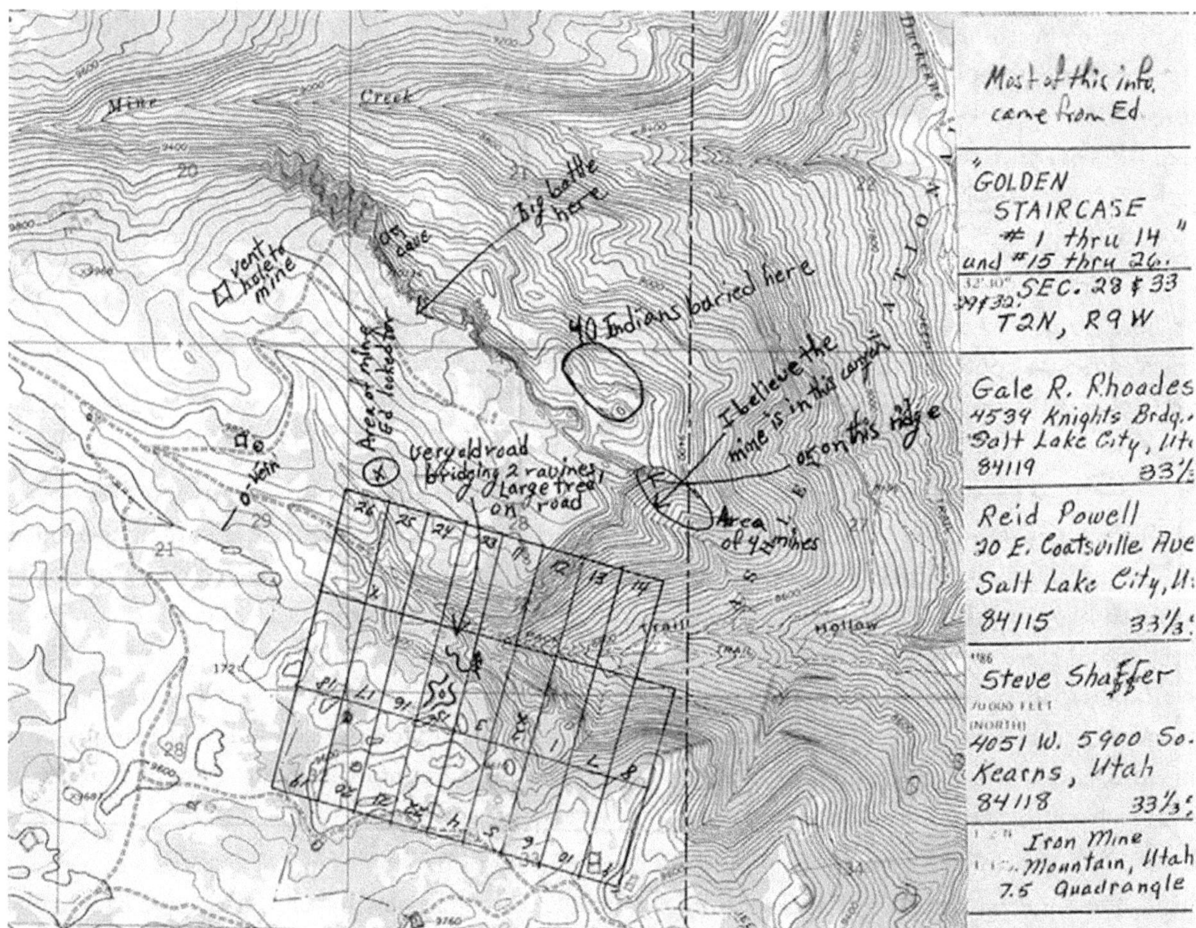

This map displays a large amount of information. Gale Rhoades and Steve Shaffer once had mining claims in Trail Hollow that they called the Golden Staircase Claims.

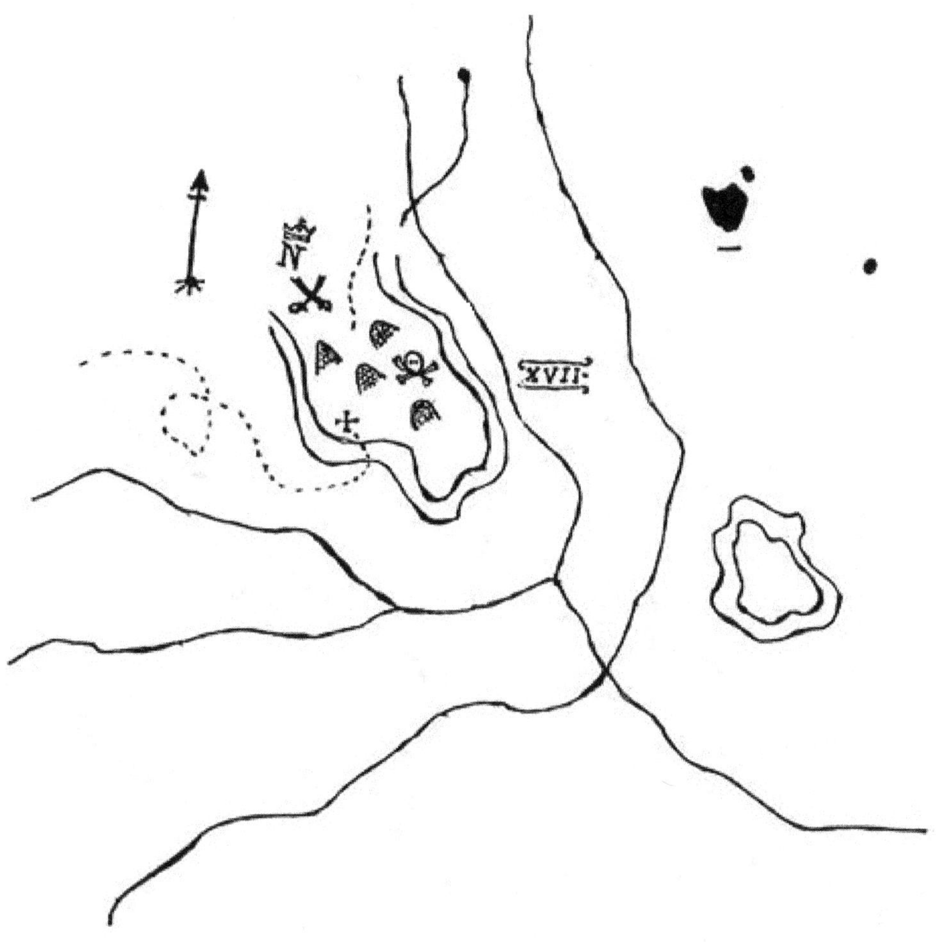

When I purchased this map from Aaron's estate, they told me it was a treasure map to a treasure buried on Lightning Ridge. They also said the original had more information on it.

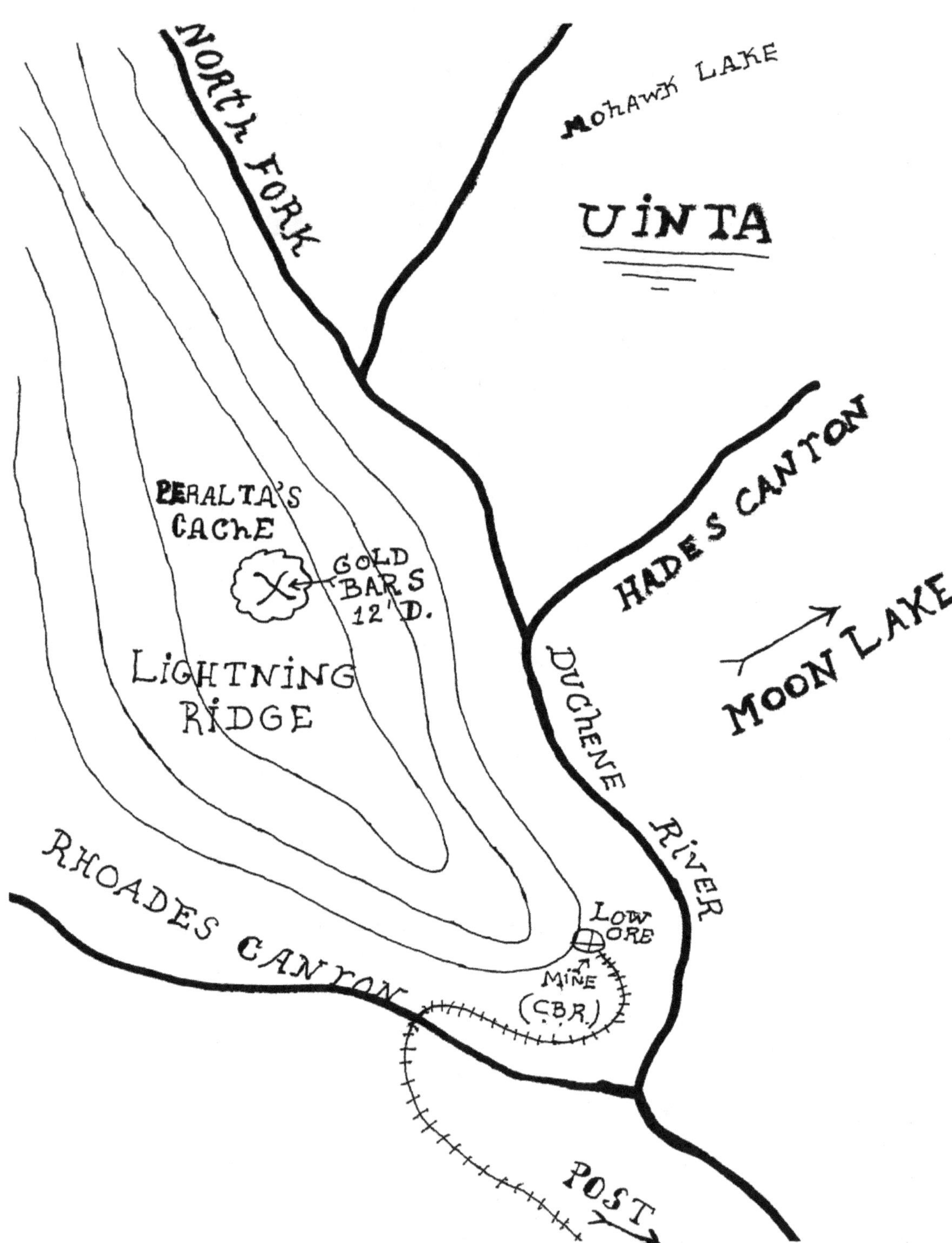

This map also pertains to a treasure on Lightning Ridge and came from the bishop.

Short Stories and Notes

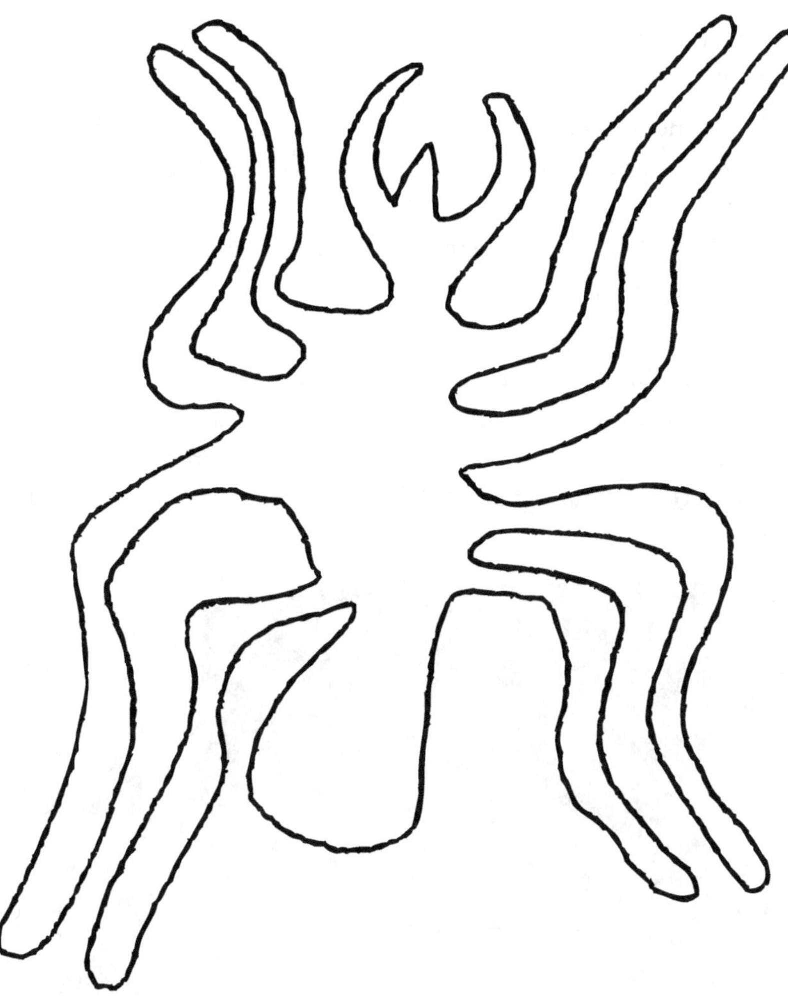

A Native American friend, who works with Cris's brother, knows of a mine with this spider chiseled into the overhanging rock of the entrance to the mine. All the Native American would say was that the mine was close to Strawberry Reservoir. I believe this mine is the same one other people have told me about, which they say is in the cliffs overlooking the reservoir at approximately 40° 18.119′ N. 111° 07.738′ W. As of this writing, I have explored the cliffs but haven't located this mine yet. The drawing above was copied from an actual rubbing. It's interesting to note the similarity between this spider and the 150-foot Nazca geoglyph spider located in Peru. They bear a striking resemblance to each other.

Massey's Rich Piece of Float

Earl Massey once lived in Dry Fork Canyon near Vernal, Utah. He has passed away. He had a keen intellect and a pleasant personality. The canyon ranch still belongs to his extended family. It's located near Sink Ridge in Dry Fork Canyon. The nearby cave carries the family name. There are a lot of fascinating things to see in this area.

Years ago my dad and I took a walk with Earl, and he told us a story as we passed a knoll and a large sink hole. His uncle, Everett Massey, and a good friend, Acel Rowley, were hiking in this area back in the 1960s when they found something quite unexpected. As they walked along, one of them noticed an unusual rock near their feet. He reached down, picked it up, and soon realized it was out of place. It was colorful, heavy, and appeared to be highly mineralized. What it was exactly, they didn't know, but they thought it might be valuable. They made the decision to send it to the Brigham Young University to be assayed. Earl said it caused quite a commotion.

Within a short time of sending it out, a man showed up at their door. He had thoroughly examined it, flown to Vernal, and asked if he could buy shares in their mining venture. You see, the sample was very good! They had to disappoint him by telling him the truth; they didn't know the source of the rock. It appeared the rock had been dropped long ago from a pack horse or by someone who had traversed the old trail they were on.

As we looked around, we could see two trails running side by side going up the side of Sink Ridge. One was old and the other one was even older. We followed one of these trails to the top, where it turned northwest, following the contours of Sink Ridge. In this area we found symbols on trees, but we lost this symbol-blazed trail after a while. We didn't find the source of the float but it sure was an interesting walk. I spoke with Earl on August 9th, 2001. He gave me permission to share this story with the public. I thanked him for it.

This knoll where the float was discovered is located on the southwest side of Sink Ridge and was probably created by movement of the hillside above long ago. The float was discovered near the base of this mound near the bottom center of this photograph. If you decide to follow the clues, best of luck in your search.

Near the bottom center of this photograph, where the hill levels off at its base, is where Everett Massey found the rich piece of float.

The Dry Fork Canyon Monuments

Author's Note

I heard the following story as I sat and visited with Larry Hacking at his Vernal home in the 1980s. It is retold in my words below.

The Hacking family had a friend named Vern who lived in northern Utah. He was a school teacher by profession and was taking a break from work when he came to see them. He had come to the area of Vernal and Dry Fork looking for a couple of rock monuments and proceeded to tell them why. Vern told them about an interesting dream he'd had. In this dream he was looking for his father. He was in a canyon, on a knoll or hill, near a couple of small rock monuments. He walked down the side of the hill calling for his father, saw an opening in the hill, and slipped inside. He walked gently down some stone steps as he entered. As he descended he noticed he was entering a room. The room was filled with artifacts. Many were made of gold and all were intricate and breathtaking in their beauty. As he gazed around in the semidarkness, he noticed a large statue in the image of a man in the room's center. Thinking it was his father, he burst out, "Dad!" Nearing the statue, his heart dropped as he realized it wasn't his father. With his adrenalin still racing, he heard voices of people entering the same way he had. He quickly looked around and located another way out. He slipped out into the bright light of day and found himself standing on the same hill he had started from.

Upon hearing this story, Larry became worried that his friend might find his secret hill filled with monuments. He quickly slipped up the canyon to dismantle them. They reminded him too much of the ones Vern was looking for, and he didn't want them to be discovered, although I never understood why. Vern's drive up the canyon was uneventful. He told Larry he didn't recognize any landmarks and returned home to northern Utah. Larry was relieved. The monuments sit in disarray on top of the hill to this very day.

This is Monument Hill in Dry Fork Canyon located at 40° 35.894′ N. 109° 43.584′ W.

The Big Spring Gold Story

Author's Note

This story comes from Mark Christensen, who heard it from a Native American. This was probably Rex Curry. Rex Curry, Elmer Iorg, Verl Iorg, and others ran around together and had solicited Mark's help in locating an old Spanish mine on Rock Creek. Their procedure went like this—they gave Mark a few clues and then sent him off to look for the mine. Upon returning unsuccessful, he would receive another clue and be sent off again. Sometimes he rode along with them, and he learned much about the mines and the history of the Native Americans and the Spanish. Before Rex died, he was going to show Mark a cave which contained many ancient artifacts, such as gold cups. Rex's plan was to take the items, sell them, and use the money to help the tribe. Rex died during the winter just prior to the summer he had planned to take Mark to the cave.

Big Spring is situated on Indian Land virtually on the banks of the Uinta River. Many years ago the Indians held their ceremonies there. Since gold was sacred to them they often used it in their rituals. They traveled to Big Spring, set up their camp, and then a few of them left to obtain "money rock" from which they would procure their sacred yellow metal. They traveled north from Big Spring and crossed the Uinta River to a certain spot known only to them. After obtaining enough for their ritual needs, they returned to Big Spring. Large fires had been built by those left at camp, and the money rock was placed directly into the hot flames. The rock was left there all night, and by morning the fires were only ambers. At that time sticks were used to pull out the clumps of their precious yellow metal. The gold was used in their ceremonies, after which it was buried or otherwise disposed of.

There is a lookout spot on the way up Pole Creek Canyon where you can park and then walk a short distance up to the crest of a small hill just north of the road. From this location you can see a long distance up the Uinta River. This is where Rex wanted to take Mark and show him something. Is it possible the secret cave can be seen from this vantage point? We may never know the answer to that question.

Woolsey's Gold Mine near Low Pass

Author's Note

Here is the James Woolsey gold mine story as it was relayed to me.

James B. Woolsey was a sheepherder all his life. He had a prospecting friend who was a dentist in Salt Lake City, Utah. They had an arrangement worked out. Woolsey would bring him rock samples from various areas where he herded sheep. If they looked good the dentist had them assayed. The dentist promised Jim that if he ever found anything good he would stake it and give him a percentage.

One particular year Jim took his herd of sheep above Current Creek just east of Low Pass on the summit. He also made his camp in that same area, just east of Low Pass. Several days later he rode his horse down over the hill, north and east of camp. He hadn't been gone long when he spotted a mine tunnel. He got down from his horse and looked at the tunnel. At the entrance, the props were sticking out of the ground. It looked as though the hillside had slid or the mine had caved in. Around the props, the rock in which the tunnel had been driven was exposed. It was a dark reddish brown in color and in

This tree, along with the one below, is a Woolsey tree in the Low Pass area. Could one of these be the right marker, or is it one we haven't found yet? The inscription on the trunk reads, "J. Woolsey 192 . . ."

the rock was a gold-colored metal that he thought might be iron. He chipped some samples from around the tunnel to take to the dentist.

During the time Jim was on the mountain, he rode by the old mine several times. That fall when he returned to Salt Lake City, Jim took the samples to the dentist's office only to find out the dentist was out of town. He left the samples at the dentist's office anyway. Within a couple of days, Jim got a job offer in New Mexico and left Utah for a time. When he returned the next year to visit the dentist, he learned that the dentist was in a mental institution. Jim found out the dentist had had the rock assayed and that it ran $57,000 per ton. He also discovered that the dentist had spent most of the winter and spring hunting for Jim until his nervous collapse. Jim did not know whether or not the ore he had brought him was the cause of his condition, but with that in mind he never went back to the old mine. Jim gave directions to find the mine and said it should be easy to find.

Here are the instructions: Go up Current Creek. Near the head is a road that takes off to the right. Follow that road up the mountain and over the top. Go east through Low Pass and continue going east along the summit. East of Low Pass you can find the camp where Jim stayed. You will know it is the place because all the trees have Woolsey carved on them. Just south and down the hill is where you can find some old dipping vats that were used to dip the sheep in. After finding Jim's camp, go a little northeast from there, down over the hill somewhere around one-quarter mile. The tunnel is in the timber, and from the tunnel you can see the west fork of the Duchesne River. Another clue that may help is to find the Bobby Duke Trail. Jim said the tunnel is visible from that trail if you look very closely. Good luck with your search and who knows, we may see each other on the mountain in our quest for gold.

The inscription on this tree says, "1918 JBW."

Payson Canyon's Cursed Mine

Very few people have ever heard this story. It happened to a very dear friend of mine named Harold Nelson of Pleasant Grove, Utah. He originally told me this story around 1985 while we both worked at Brigham Young University. He retold it to me November 1, 2001, and gave me his permission to put it in writing on my website and in a book (if I ever wrote one). Before telling this story, I wish to thank him deeply.

This story takes place near Payson, Utah. In the late '50s, Harold drove trucks for Highland Dairy. He delivered milk to various stores in the South Utah County region. One day while at the Spring Lake store, he waited in line behind an elderly gentleman. Harold noticed that when the elderly man was given his bill, he went behind a food stand and pulled out his shirttail, where his money was pinned, and retrieved cash to pay for his food. After paying for his groceries, he seemed to just wait around. When Harold was done with his business, the gentleman asked him for a ride to Payson. Harold said, "Sure, I can do that." What came next was quite peculiar. He asked Harold to drop off his food at his house and to place it behind the rose bushes. He then asked Harold to drop him off a couple of blocks away. Harold thought this to be very odd since driving a very large dairy truck to a house would be highly visible. However, he did as requested.

Over the next little while, Harold and the gentleman became friends. Harold delivered to the Spring Lake store on Thursdays, and the gentleman was usually there buying his food. Even though this man appeared to be in his eighties, he walked everywhere he went. Harold often saw him walking from Payson to Spanish Fork and gave him a ride. The gentleman seemed to be very secretive and didn't trust anyone except Harold. One day he asked Harold if he ever drove up Payson Canyon. Harold replied that yes, he did, and explained that he liked to go fishing at Payson Lakes. The gentleman then asked if he could drop him off at a certain spot up there next time he went fishing. "Sure, I would be glad to," Harold said.

This is a photograph of the bridge where Harold dropped off the old prospector. Where he went from here is anyone's guess.

When Saturday came Harold was ready to go fishing. He picked up the gentleman and took him up the canyon. He dropped him off at a bridge not far above the Grotto, and the gentleman offered no explanation. That afternoon, as Harold was returning down the canyon he stopped at the same bridge and honked. The gentleman came out of the trees and hopped in. Harold did this several times. Finally, on one of these trips the gentleman confided in Harold that he was looking for a gold mine. It was a very rich mine which he had found and lost. He said he had marked the trees with his ax while going from the mine to the road, but upon his return he found that all of the trees were marked. Because of this, he said it must be Brigham Young's cursed mine but he wanted to find it again anyway. He had been there, seen the gold, and strongly wanted to return to it.

He asked Harold for money to grubstake him (finance the mine in exchange for a portion of the profits) so he could look for the mine full-time. Harold didn't have the financial means to do that. The gentleman was only asking for $20 per month, but it wasn't to be. He indicated that Harold was a good, honest man and that the powers that be might smile upon both of them with that arrangement. Harold said, "I'm sorry, but I don't have the extra money to spare."

Harold moved to Pleasant Grove in the early '60s and lost track of this gentleman but never forgot him. There isn't a time when he goes up the canyon that he doesn't ponder about his old friend. As he drives past the spot where he dropped him off, he wonders about the mine, what it's like, and where it is. If you choose to go looking for this mine in Payson Canyon, remember to look for the trees with the marks on them. The bridge (drop off location) is at 39° 56.642' N. 111° 40.698' W., elevation 6811 feet. Yes indeed, the mountains do hold mysteries of riches. But then, that is the stuff dreams are made of.

The Burro Mines

The Burro Mines do exist. A close friend of mine took me to them in 2001. You can still see the depressions where the mines were. The symbols are there and they are exciting to see. When I view pieces from our past like this, I can't help but wonder how much longer they will be with us.

I promised my friend I wouldn't divulge this location to anyone. Maybe by so doing, these trees will be here for future generations to stumble upon.

This picture shows one of the symbol trees near the Burro Mines.

A Spanish Silver Bar

The pictured gold finger bar is a replica and is similar in looks to the real finger bars found at the salvage site of the sunken Spanish Galleon, the Nuestra Senora de Atocha. The silver bar, however, is real. A friend found it on Hoyt Peak long ago. In March of 2002, I had access to it and was able to photograph it. The bar measures about nine and one half inches long. The owner had its content checked and said it was eighty-six percent silver, thirteen percent gold, and the balance is copper. Through conversation, study, and research, I learned several things about the silver bar.

1. The owner once turned down $10,000 for it.
2. Its markings match that of gold bars found at the recovery site of the sunken Spanish Galleon, the Nuestra Senora de Atocha. It is marked as being ten and one quarter carat pure (twenty-four carats being pure gold). The markers that indicate this are the X and the dot. Compare the markings on the replica bar which are twenty-one and three quarter carats. The silver bars at the Atocha site and the bar pictured have similarities in their owner marks and shipping marks, but the silver bars found at the Atocha site didn't have purity markings on them. The gold bars did.
3. The symbol at the left end of the silver bar is most likely the owner's mark.
4. The odd symbol just left of center is probably the shipper's mark.
5. Each letter in the word HISPAN, stamped several places on the bar, appear backwards. However, you will only notice it in the stamping of the S, P, and N. The manufacturing of the dye for the stamping could have been made by someone who was dyslexic. Some of the cob coins found at the Atocha site were the same way. An accurrate reference for information on these bars is *Cob Coinage*.

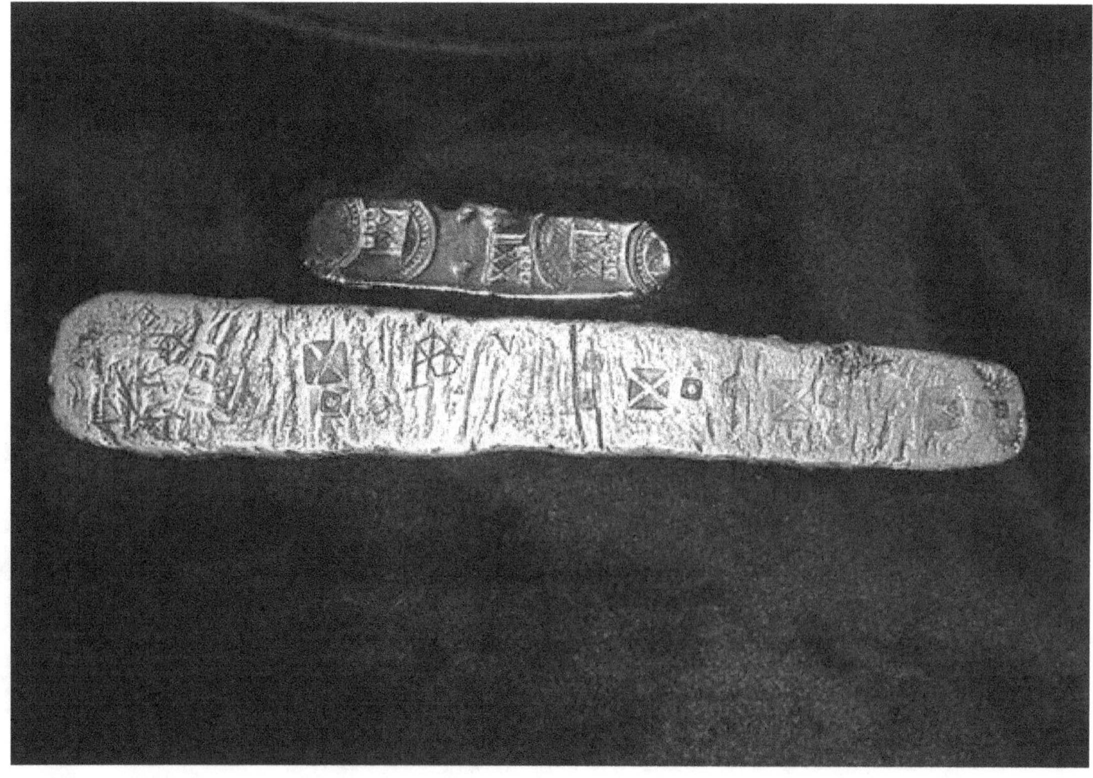

The Ed Twitchell Connection

On February 21, 1987, it was my treasured privilege to meet and interview Ed Twitchell. Ed and I never hiked the hills together, but shared many memories through phone conversations, visits, and letters. He had a childlike attitude about these adventures. To study the legends and learn about the clues hit the spot for him. It is probably the highest tribute I can give him to say he shaped my attitude and outlook with his. Although he is no longer with us, I will always feel close to him.

I came to know Ed through his daughter and son-in-law, Vicky and Lynn. Lynn told me about a job opening at Brigham Young University, which I applied for and received. When Lynn discovered that I enjoyed the study of the Rhoades mines, he said, "You should meet my father-in-law. You would probably hit it off." As they say, the rest is history.

Ed was a unique individual. He grew up in the basin. He lived near a full-blooded and highly informed Native American named Richard Ridley. Ed also knew other Native Americans such as Cump Murray and Don Foote, to mention just a couple. He later enlisted in the military with Richard and actually worked with him at the military depot. Because these Native Americans were so active and so closely involved with the Native American affairs, they knew far more than most people realized. Much of this information they passed on to Ed and, in turn, was passed to me.

When it came to Ed's knowledge, it was mainly related to the matters pertaining to Native American Land. However, he also knew about many other things and had some very interesting experiences. He was always very sensitive to others' feelings, and I hope I will not disgrace him with anything I convey here. It's my hope that I can pass on his information and experience so that it will not be lost.

Ed told me many stories and important clues. I will try to retell this information as accurately as possible. Ed encouraged me to walk the length of the South Flank fault. He said I would find many rich mines by so doing. I have only followed his advice to a small degree, but have no doubt he is 100 percent accurate. It hasn't come to the top of my priority list yet.

Cump Murray's Rich Mine

This is a story Ed told me in 1987 and that I feel is appropriate here. I learned as much from this story about Ed's warm and caring personality as I did about the mine, its history, and its location.

This basic story is found on page 122 of *Lost Gold of the Uintah, The Rest of the Story*. However, no mention is made about the third person involved in this adventure. It will be told here from Ed's point of view. Ed was the third person on this trip.

The story begins as Ed Twitchell and Jack Spencer pulled up to Cump Murray's house. This outing had been planned for some time, but there was a small wrinkle that would end up turning against them. Cump wasn't feeling well that day. But try as they might to talk him out of leading the expedition, he wouldn't hear of it. He climbed into the truck and off they went.

As they traveled north, they came to the divide where the lower road leads to Yellowstone and Swift Creek and the upper road leads to the head of Hells Canyon and beyond. They took the upper route and a short distance past the split, reached the well-known deer hunter's campground. Here they veered to the left and abandoned the better road altogether. Now the road became steep and rough. At one point Ed asked Jack and Cump to hop out and move some rocks from their path. Jack hopped back in, and Cump got in as well but much slower than expected. They started up again, when Ed noticed Cump had slumped over. Ed asked Jack to check on him, and upon doing so, he realized Cump had left this mortal existence.

Ed told me it was the most difficult thing he had ever done in his life to take Cump back to his family and tell them what had happened. I could tell Ed had great love for his friend Cump and would miss his friendship greatly.

Ed said he tried to find Cump's mine for two years after that but didn't know exactly where to look. Cump had told him the mine was three miles from the campground. The rough road was one and one-half miles long. They were planning to drive to the road's end, get out of the truck, and then hike one and one-half miles to reach the mine. The present road isn't the same as it was back then. The present road system is much more extensive. If someone were to look for this mine now, they may have to examine an older map and note the differences.

They had picks and shovels with them and were planning to uncover the mine to reach their goal. Cump told them he had made a withdrawal from the mine in the interim between his uncle showing him the mine before his uncle died and this outing with Ed and Jack. Cump's uncle also told him he had once had to escort some Spaniards away from the mine and warn them never to return.

Yes, this truly was a rich mine. Cump said the material he had pulled from the mine assayed at twenty-seven ounces per ton in gold and forty ounces per ton in silver. This was a time of frustration and sadness for Ed to lose a dream and a friend both on the same day.

The Crescent Mine and Tworoose Pass

Ed told me many stories about Cump and his adventures in the hills. I enjoyed retelling these and other stories to my children while tucking them into bed just like my father did for me. Here are a couple in writing so they are not lost or forgotten.

Cump knew many locations where gold could be, shall we say, obtained. Whether it was by mining, scooping, or just picking it off the ground, it didn't matter to him. He knew the locations like the back of his hand, and when it fit his fancy, he plucked it from its hiding spot. Cump and gold seemed drawn to each other like hardened steel to a strong magnet.

On one particular outing, Tworoose and Don Foote accompanied Cump. Don was assigned to watch the horses and wasn't allowed to follow them to the mine. Cump and Tworoose went to a rich mine inside a hill in Brown Duck Basin. This mine had two entrances, and they spent time, slightly at odds, discussing which entrance they would enter. The north entrance was smaller, which didn't accommodate Cump very well, and the western opening was rocked up, which would take longer to reopen but would be larger once the walled-up rocks were removed.

This mine was called the Crescent Mine because of the markers set up in a crescent moon shape around the hill's base perimeter from the northern to the western side. It was large inside, having three levels, and was even purported to have two Spanish cannons enclosed. Some of the gold inside is said to already be bagged up and ready to haul. The gold in this mine is said to be a beautiful butter yellow color in white quartz.

They eventually made it inside and procured the amount of gold that would fill their needs. Then they met back up with Don. Don never knew the gold's source this time or at another instance which I will tell you about now.

This location wasn't very far from the setting of the last story. Don and Cump were camping near Tworoose Pass. One morning when Don awoke, he found that Cump wasn't in camp. However, before long, Cump came walking into camp. He held out his wet, clenched hand and slowly opened it for Don to view. Don was amazed to see several gold nuggets, the largest being about the size of a small chicken egg. Cump wouldn't divulge the source of the gold, but Don guessed it was placer gold from a watery source near camp.

Along with the gold, Cump was carrying a piece of bark that had been fashioned into a long spoon. Don imagined it was used as a tool to dip deeper into a pool or stream than someone normally could reach, and that Cump used it to pluck the nuggets from their hiding. If only Don could retrace Cump's footsteps. Then he would know one of Cump's secret locations and pluck some gold for himself.

As far as we know, Don never discovered any of Cump's hot spots. When Cump passed away, he took many secrets with him.

If you go up there looking for that silver mine, there or 2 or 3 things I forgot to tell you. A man made a lifeng paning the Gold & Silver on swift creek. What you look for is Black Rock If you took you pan & paned the River look for Black or Black Brown Rock. If you can find the Black Rock & you go up the River & you can not find any more you have went past It. good Luck Ed-

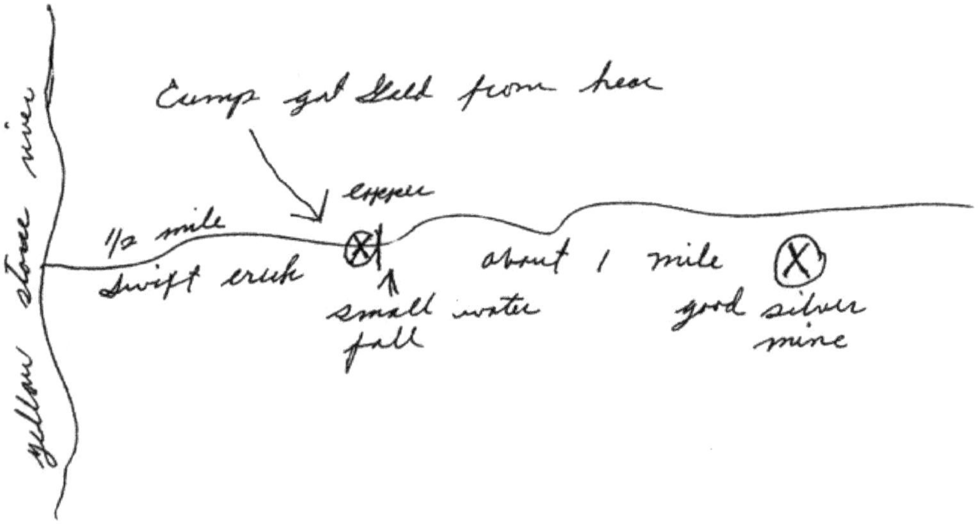

Camp got Gold from hear

copper

1/2 mile

Swift creek

small water fall

about 1 mile

good silver mine

yellow stone river

This is a map from Ed and displays Swift Creek and the mines there. The explanation of this map follows on the next page.

Following Ed's directions, I was able to locate this colorful and unique rock formation on Swift Creek. Even with my untrained eye, I could tell something geologically significant had happened here.

However, Cump's nugget prospect wasn't obvious to me. I believe the water has changed course over the years and has covered up any trace of Cump's diggings. My son Del Ray and I panned but didn't find any gold, and we were a little uncertain exactly where to look.

When Ed later saw these photographs, he laughed and said we were in the right place. He told us Cump Murray dug gold from the side of the stream we were on, near the water's edge, and at the tail end of this massive outcropping of copper colored rock.

The Forest Service maps of 1980 show this area as Forest land. However, the maps of 1984 designate this location as wilderness. The government changed the boundary line and gave us a new challenge. We will follow our dreams to another location.

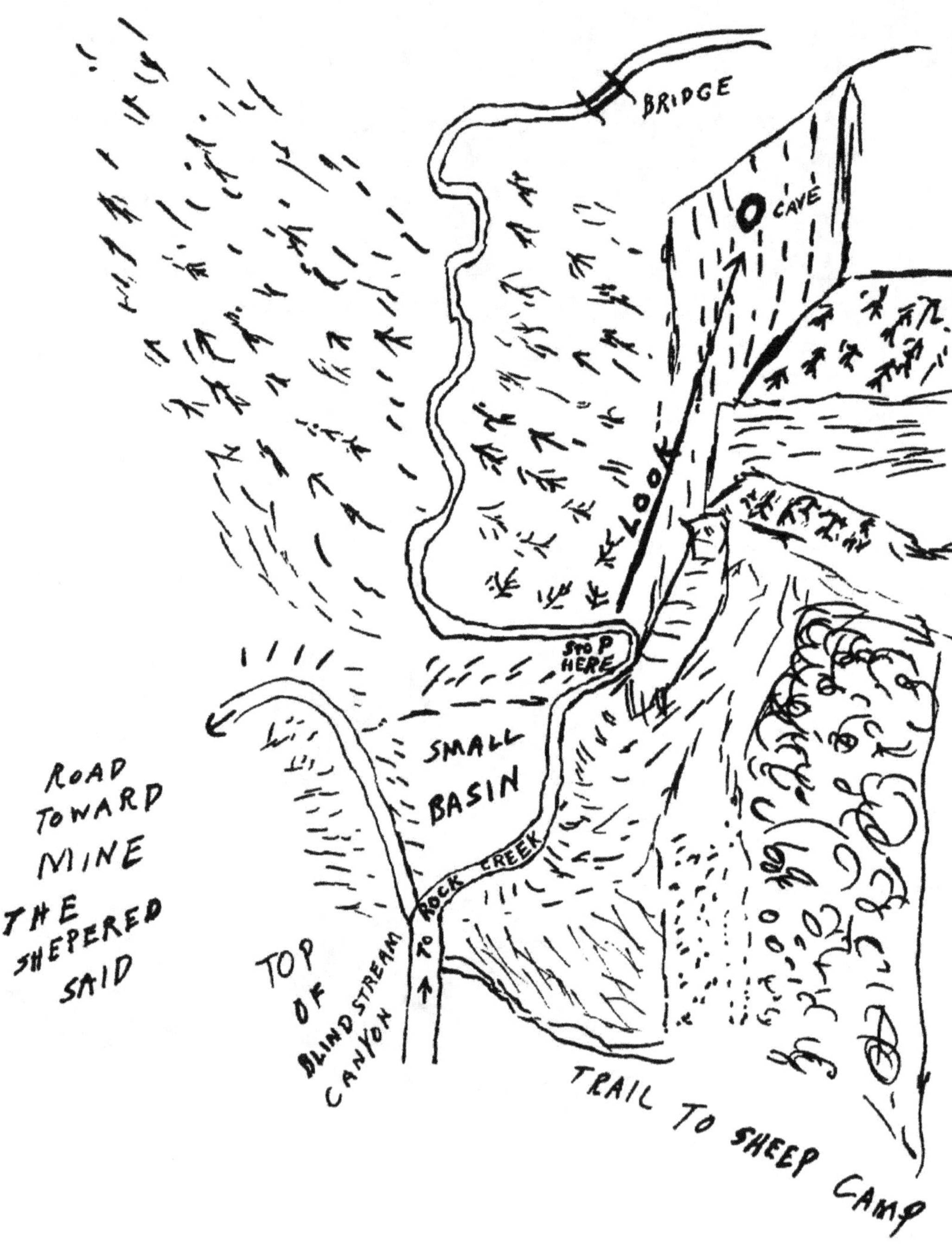

This is a map drawn by a Native American for Ed. He didn't tell Ed what was in the cave but said that there was something important inside. Because it is a difficult place to access at best, Ed never checked it out. He passed this information on to me in hopes that I will someday traverse this rugged terrain and enter this mystery cave.

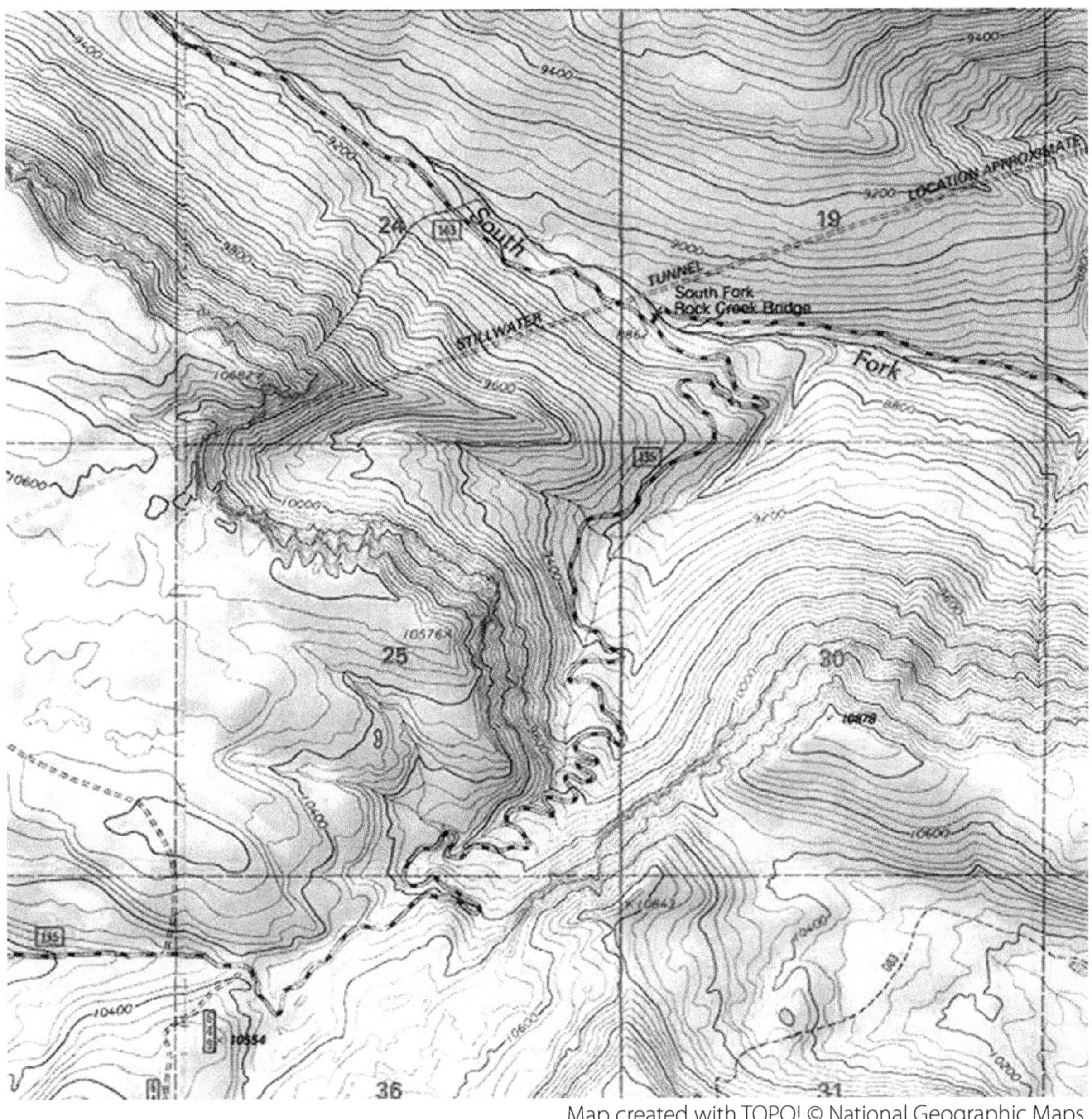

Map created with TOPO! © National Geographic Maps.

Here is a map of the rugged hillside in question. I have noticed several caves on it. The caves have a draw for me, but I haven't yet tackled these treacherous ledges and loose rock hillsides.

I recently read through many of Gale Rhoades's notes and documents in preparation for this writing. Gale states time and again how dangerous this particular hillside was for his family as they attempted to explore its alluring caves.

Gold Hill

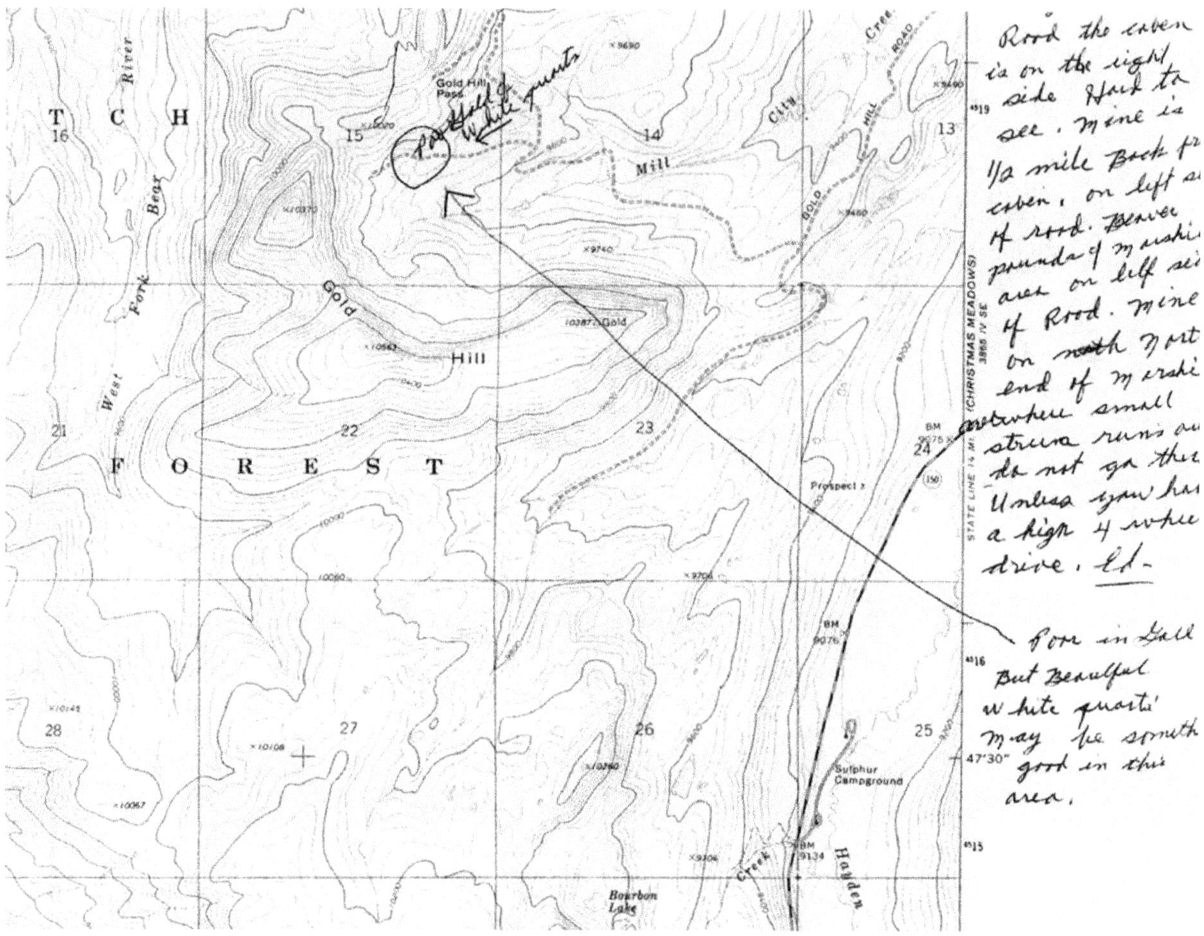

Here is some information Ed wrote on a map for me. It pertains to the Gold Hill area of the Uinta Mountains. It is written out below and slightly altered for easier reading.

> *The cabin is on the right side of the road and is hard to see. The mine is one-half mile back from the cabin, on the left side of the road. There are Beaver ponds and a marshy area on the left side of the road. The mine is on the north end of the marshy area where a small stream runs out. Do not go there unless you have a high 4-wheel drive. Ed*

The arrow to the circled area says, "Poor in gold but beautiful white quartz. There may be something good in this area."

Are There Emeralds Too?

When I think about the shipments of gold and silver that left the Americas for Spain, I also picture precious stones such as emeralds being on board. At the salvage site of the sunken Spanish galleon the *Nuestra Senora de Atocha*, more than four hundred emeralds have been uncovered near the main ballast pile. It appears the Spanish craved precious stones almost as much as precious metals.

Many treasure hunters will not pursue a treasure unless they have at least three independent sources that corroborate each other. Such action is wise. Therefore, I will cite three different and independent sources that verify the existence of emeralds in the Uinta Mountains. The following information is about a naturally occurring, obscure treasure.

Source #1

In the mid '90s, Mark Christensen and I visited a man in Roosevelt, Utah, by the name of Gordon DeLapp. We were there because we had heard he had discovered an emerald in Utah. We were on pins and needles because this was the first we had heard of such an occurrence in our home state. As we visited with him on his front lawn, the conversation eventually got around to the subject of his find. He stated that he had been gathering rocks from various locations close to his house and had placed his collection in a bucket. One day he was going through his collection and noticed a fist-sized chunk of quartz. He said it was rectangular in shape and both sides were flat except for a square-looking bump on one side. He took the quartz, set it on its side, and struck the opposite edge with the claws of his hammer. It split open, exposing a deep green, breathtaking stone. He had suspicions of what it might be but took it to a jeweler to be certain. The jeweler confirmed his hunch and weighed it. I don't recall that Mr. DeLapp told us its weight, but he said it was about three-fourths of an inch long, one-half inch wide, and one-half inch thick. He showed us a picture and said the stone was at the local bank for safekeeping. Looking at the photograph, we couldn't believe our eyes. Talk about beautiful! It was a mesmerizing deep green color.

We finally got down to discussing the source of the stone. He said he couldn't be absolutely certain but believed it came from the south fork of Rock Creek. He further described the spot. As you travel toward Arta Lake, he said, you cross a bridge and are then going northeast. Within one-quarter mile of crossing this bridge, you will see a spring on the left side of the road and at times the water seeps across the road. The spring is associated with the south flank fault, which is exposed at this location. Mr. DeLapp said the quartz probably came from the dirt at this spring.

Mark and I thanked him for his time and made our way up the canyon to check it out for ourselves. We took some of the material from the spring and panned it, with the idea in mind that emeralds would be heavier than most other material encountered. Each of us panned a couple of pans of dirt and rock but didn't come up with anything good, even viewed under a magnifying lens. We will be back again and check some more.

This is a picture of the emerald discovery location on the south fork of Rock Creek. Here are the GPS coordinates to it: 40° 33.879′ N. 110° 46.297′ W., elevation 9474 feet.

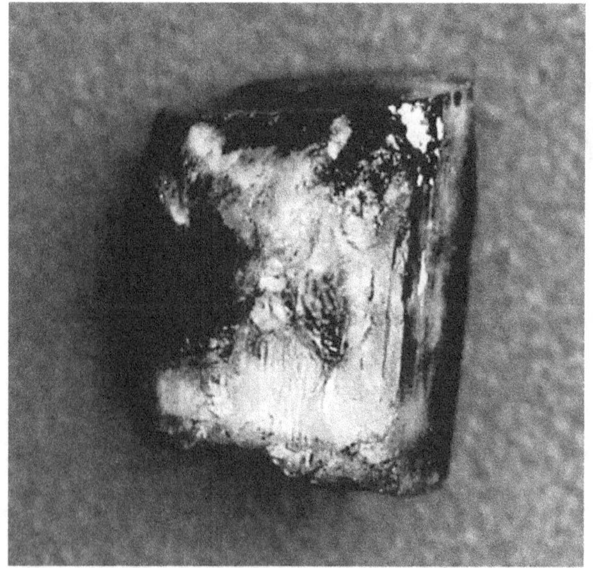

Even with the matrix still attached, the pictures above can't hide the beauty of Gordon DeLapp's emerald. It measures nearly an inch in length and weighs approximately 30 carats.

Source #2

The following story is a brief account from Jeffrey Keith and Timothy Thompson. Jeff is a Brigham Young University professor and geology department chair. He is the individual who first told me about the emerald found by Gordon DeLapp.

Tim was contacted in 1995 by Sheila Ivers, who ran a retail jewelry store on 3300 South and Highland Drive in Salt Lake City, Utah. She said that a Mr. Gordon DeLapp had dropped off a deep green stone for identification. She asked Tim for assistance. He picked up the stone and skeptically looked it over. He was already doing microprobe work on red beryl as a thesis and decided that was the surest way of determining the stone's makeup. The tests were performed and confirmation given that it was indeed an emerald. Tim conveyed his findings to her. Sheila had told Gordon that if it were an emerald, they would form a small company and go looking for more; after all, she was in the jewelry business. Between the time they determined it was an emerald and Sheila officially made a deal with Gordon, he decided to go on his own. He decided to go up and stake claims, even though it was winter. Sheila got in touch with Floyd Anderson and others who were interested in investing and forming a company. The first time the group met, they discovered what Gordon had done. Because they didn't know where Gordon had staked his claims in the snow, or even exactly where he had found the emerald, they decided to wait for spring and then take a look.

It was June 22, 1996, when Jeff and Tim made it up the south fork of Rock Creek. They hit snowdrifts and had to walk some distance to the discovery location. They passed Gordon, who had gone up looking for more emeralds. Tim and Jeff looked at the rock formations as they hiked. As they crossed Rock Creek, they noticed red pine shale, iron stains, and pyrite. Then they reached the South Flank Fault. They sized it up. There was pyrite and iron stain there too. They looked at each other in amazement as they realized this was identical to the geologic setting for the emeralds in Colombia! They were familiar with Terry Ottaway's master's thesis at the University of Toronto on the setting of Colombian emeralds. As they studied the area, they determined this wasn't a small or localized intrusive dike but rather arkosic arenite units. All of the necessary elements were here to form emeralds. They concluded that it could possibly be one of only two places like it in the world.

They also determined that in his haste Gordon had staked directly over two active claims owned by Darrel North, making Gordon's claims invalid. Darrel had kept them active because there were small amounts of gold on his claims. As time went on, Gordon was without a hold on the ground while Tim and Jeff continued to examine the possibilities. Two more emeralds were discovered in the fragments Gordon had collected. One was three millimeters in length and the other was eight millimeters. As they worked in the area, they also discovered a large deposit of commercial-grade fibrous calcite. It was south of Darrel's claims and turned out to be a viable discovery, but that's a story for another day.

Source #3

The hot-off-the-press book of November 2005 was entitled *Uinta Mountain Geology*. In the article "Potential Economic Mineralization in the Uinta Mountains, Northeastern Utah" by G. R. Conn of Duchesne, Utah, there is a subheading called "Emeralds." Conn states that "there are two regions in the Uinta Mountains that may produce gem quality beryl (emerald, morganite, aquamarine, etc.): the Red Creek Quartzite outcrop belt and the South Flank fault zone (figure 1, C and D). Many investigators of the Red Creek Quartzite have noted the presence of beryl (Bullock, 1981) and this region is geologically favorable for schist-hosted emerald deposits."

The Scientific Explanation for the Emeralds

While Jeff Keith and Tim Thompson were busy learning about the occurrence of calcite and emeralds in the Uinta Mountains, Jeff requested the help of a few colleagues at BYU. The following is a write-up by Jeff, and portions of this information were used in an abstract about that occurrence on page 385 of Uinta Mountain Geology. *Those who assisted him are listed here.*

The Genesis of Fibrous Calcite and Shale-hosted Emerald in a Non-magmatic Hydrothermal System, Uinta Moutains, Utah.

Keith, Jeffrey D.; Nelson, Stephen T.; Thompson, Timothy J.; Dorais, Michael J.; Duerichen, Erin; Olcott, Jay; Dept. of Geol., Brigham Young University, Provo, UT, 84602; and Constenius, Kurt N.; Dept. of Geosciences, University of Arizona, Tucson, AZ., 85721-0077.

Large veins of fibrous, translucent calcite, up to 0.5 km long and 20 m wide, occur in Mississippian carbonate units near the South Flank Fault zone of the Uinta Mountains. Directly underlying these veins are altered zones of Proterozoic (950 Ma) Red Pine Shale. Within the fault zone and shale, three emeralds have been recovered. The largest emerald is a hexagonal, prismatic, 2 cm, 30 carat crystal that exhibits zonal variations in color that correspond to measured variations in Cr and V content.

The host rock for the emeralds is a 1 km thick, black (organic-bearing) shale sequence with arkosic arenite units. Abundant pyrite, barite, vein quartz, and fibrous calcite veins, bleached shale and green mica are present along the fault zone. There are no igneous intrusions within 50 km of the area and no geophysical evidence to indicate that the alteration is associated with concealed igneous activity. The isotopic composition of pyrite from the South Flank fault zone (delta 34S, +4.6 to 4.2 per mil) is not characteristic of an igneous source but is similar to the isotopic signature of sulfur from some Tertiary oil field brines (about +5 per mil, Wasatch Formation and lower Green River Formation) present at the base of the Uinta Basin. The deepest portion of the asymmetric Uinta Basin is bounded by normal and reverse faults related to uplift of the Uinta Mountains. Consequently, the source of hydrothermal fluid responsible for alteration of the Red Pine Shale may be similar to that of Columbian emeralds where basinal brines (Tertiary or older) migrated to the fault margins of the basin. These data suggest that basinal brine may have mixed with other water along the fault zone or reacted with feldspar and/or dolomite to form sparse emerald and other alteration products.

Analyses of C and O isotopes help constrain the temperature and composition of the mineralizing fluids and indicate a temperature in the range of 200 to 300 degrees C from a high ionic strength fluid similar to oil field brine. As sulfate in the brine was reduced by reaction with organic carbon in the shale, a carbon dioxide-rich fluid was produced. The acidic fluid dissolved portions of the overlying limestone, and intermittent loss of carbon dioxide caused precipitation of fibrous calcite.

Natural or Altered?

The following rock formations are up for consideration as to their meaning and authenticity. Several authors have written about Spanish explorers and miners stacking, carving, or otherwise altering rocks so that later expeditions would have waypoints to follow. The list of authors may include, but is not limited to, Charles Kenworthy Jr. and Mike "Hawkeye" Pickett. Some of the clues are very subtle, while others are hard to miss.

Some people see the outline of a Native American chief with a full headdress flowing behind him while looking at this rock. What do you see?

I have heard stories of rocks carved or altered to resemble animals or other things. I didn't believe in such things until one day, while checking out a story, I stumbled upon this odd-looking lizard. I hiked up to it and noted it even had a hole tunneled out to make an eye (pictured below). It quickly made me a believer. It's in Brownie Canyon at 40° 38.264' N. 109° 46.292' W., elevation 7985 feet.

This outcropping of rock is called the Guardian. It appears to have several faces. The one at the top right is looking directly at a mine on Native American property that contains gold, silver, and arsenic. This area now appears to have sensors that may be tripped by trespassers. The caved-in entrance to this mine is pictured below. Sorry, no GPS coordinates will be given this time for obvious reasons. Remember, it's on Native American land!

I got out my camera and took this picture to keep the experience fresh in my mind and to document our find.

One day in the summer of 2002, Verl Iorg and I were exploring areas in Blind Stream Canyon. We were looking at a particular hillside when I happened to look a little bit closer at the cliff face to our right. What I saw then took my breath away.

I was beside myself with excitement the first time I saw the thunderbird or eagle shown in Gale's book. Ever since I saw the photograph in his book, I have had a strong desire to see it for myself. Ed once told me that if I found the eagle, there would be a star formed by stones placed on the hilltop above it. As of this writing, I have not yet checked for the star. I believe the star and eagle could be from a culture previous to the Spanish influence in Utah, but I lack the evidence to elaborate at this time.

Is there a correlation between the Killum Map with the star and the eagle pictured above? Until compiling this work I thought the likeness was coincidental. Now I am convinced that the three fit together like puzzle pieces. The only thing lacking is the opportunity to confirm my hunch.

While out exploring in Utah's west desert on November 11, 2005, a friend and I came upon this rock formation called the Great Stone Face. It is some miles southwest of Delta, Utah. The formation, it is said, resembles the LDS prophet Joseph Smith. The face is looking to the right in a southwesterly direction.

At the base of the Great Stone Face is this unusual-looking formation which I will call a turtle head. We didn't take the time to see where it might lead but thought it interesting anyway. The eye has been chiseled or pecked into the rock. The head is facing almost directly south. Turtles were often used by Spanish miners to indicate the direction to a rich mine or treasure.

This facial outline can be seen on the way to Mosby Mountain. It's located near Mosby Creek. This photograph was taken during the summer of 2005 in the evening hours, when the face can best be seen.

My brother-in-law, Ken Lance, is standing on the formation for this photograph.

This is the place the Native Americans call Towats Point. It is said that the word *towats* means man or man-like in one of the Native American languages. It is most easily seen during the morning hours. The face is looking west overlooking Whiterocks Canyon.

This unusual-looking face is found on the Duchesne River. The face is seen here from two different angles.

What is your opinion? Are these formations natural or altered?

Photograph courtesy of Robert King

This remarkable image shows a shape resembling a face. (Hint: it's just right of the center.) Notice that one eye appears open while the other is closed. It even appears to have teeth. It's in Squaw Basin just west of Cleveland Peak. This photo is courtesy of Randy Lewis.

On July 4, 2006, while exploring with some friends near Lightning Ridge, we noticed this unusual rock formation. My friend Michael Mansfield said it looked like a turtle, and I totally agree with him. In fact, it appears to be looking directly at us. Mike Pickett often speaks about turtles and their significance. Turtles are directional markers.

As we have explored and studied the symbols, my estimation of turtles has been that they are very important and should be followed. They don't appear to have the same significance as eagles and thunderbirds, or they may be from a different time period in Spanish exploration and mining. Although they may not be as important as eagles, they do mean you are on the trail to wealth in the form of a mine or treasure. Therefore, it may be financially worth someone's time and energy to follow them to their destination.

There are plenty of surprises still to find. On April 28, 2006, while on a hike in Utah's west desert, I stumbled upon this sleeping lion. Its meaning is yet undetermined. Whether it's natural or not is beyond my ability to say. The resemblance is obvious and can easily be seen for a considerable distance.

For anyone interested in seeing it in person, you need only go for a hike up Miller Canyon on Sawtooth Mountain in Utah's House Range. It's up high on the south side of Miller Canyon. If you decipher its meaning, I would like to hear your explanation.

Maps

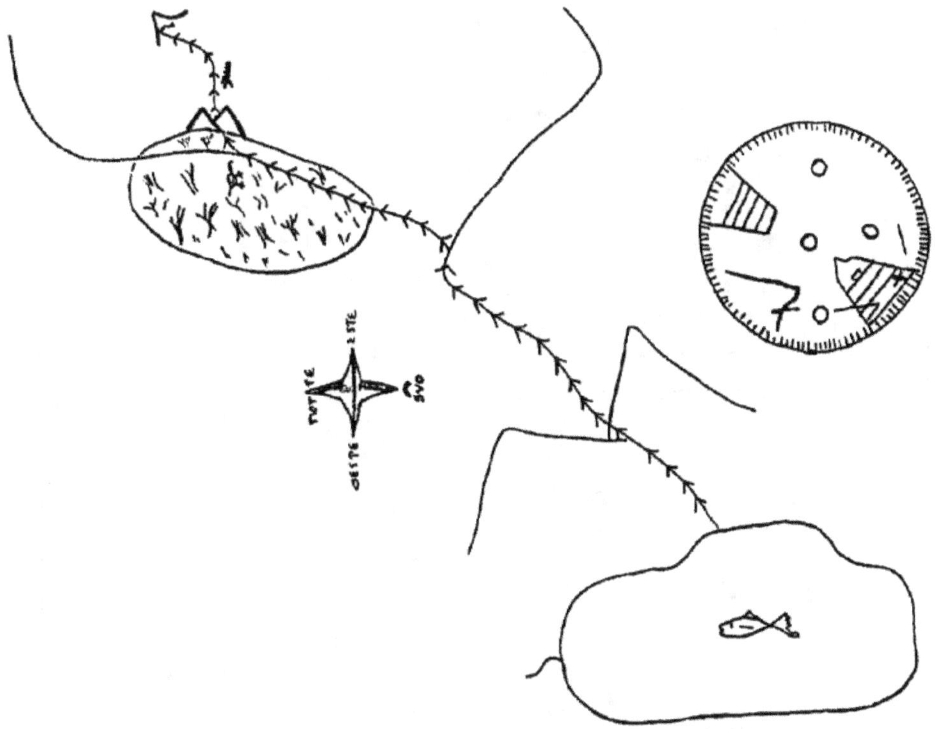

Many years ago, Ed Twitchell gave me a copy of this map, and in tracking its origins I discovered it had been drawn by June Balaich. This map appears to be geographically correct, showing Utah Lake, Provo Canyon, and a mountain that may be Hoyt Peak. According to June, the circular object shown on this map represents a map made from a bone. The map was found in a mine above Wallsburg by a family named Woodward. According to Aaron, a series of altered trees lead the way to Hoyt Peak. He described just such a tree on Hoyt Peak. He stated it had fallen. He also said that according to the Native Americans, there was a golden Christus buried in the vicinity. It would be interesting to find that tree if it still exists.

Upon determining where the map came from, I called June and he gave me permission to include it in my book. It has been copyrighted by June, and I use it here with his consent.

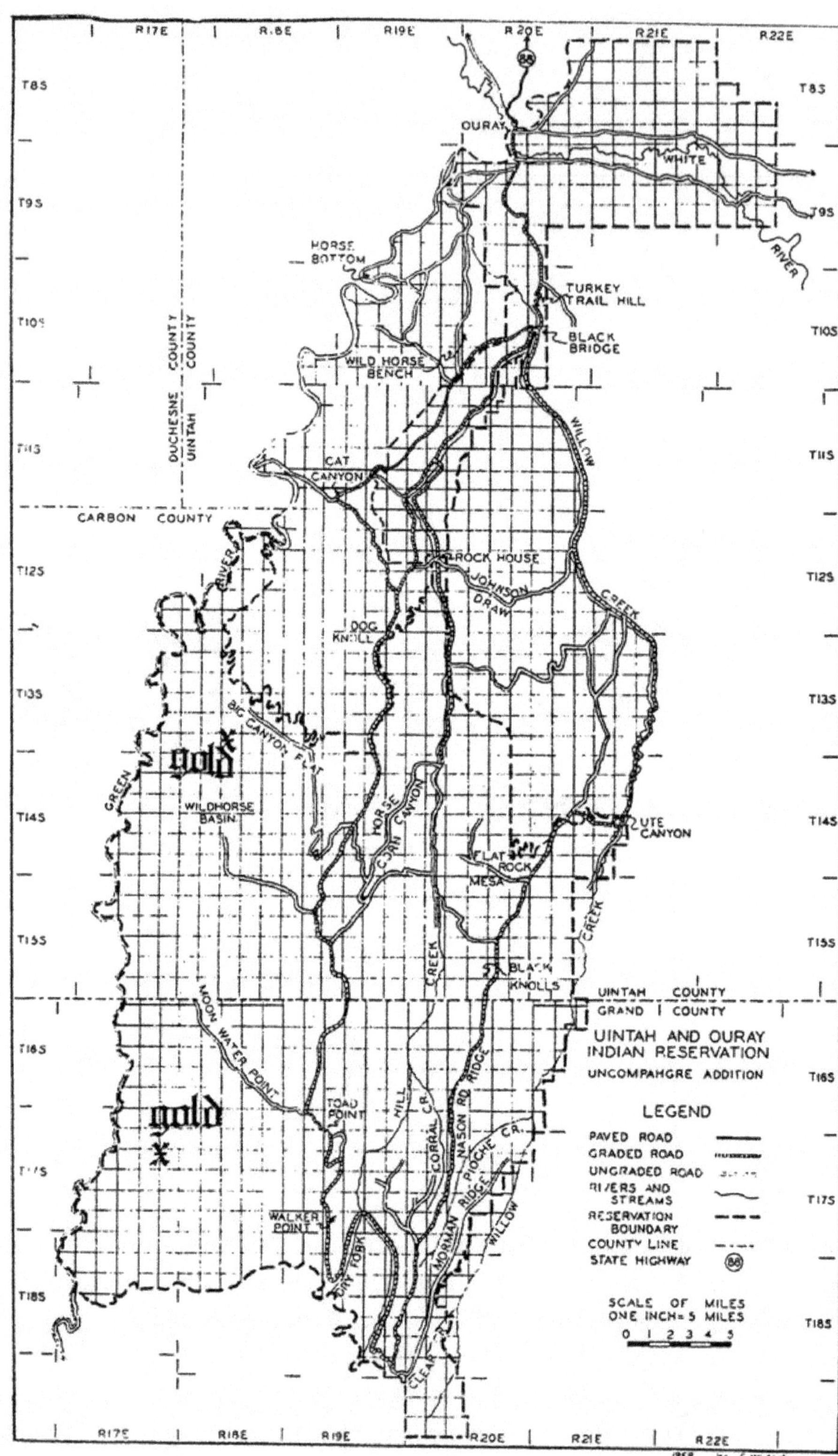

This map shows the Native American extension land and two locations of gold.

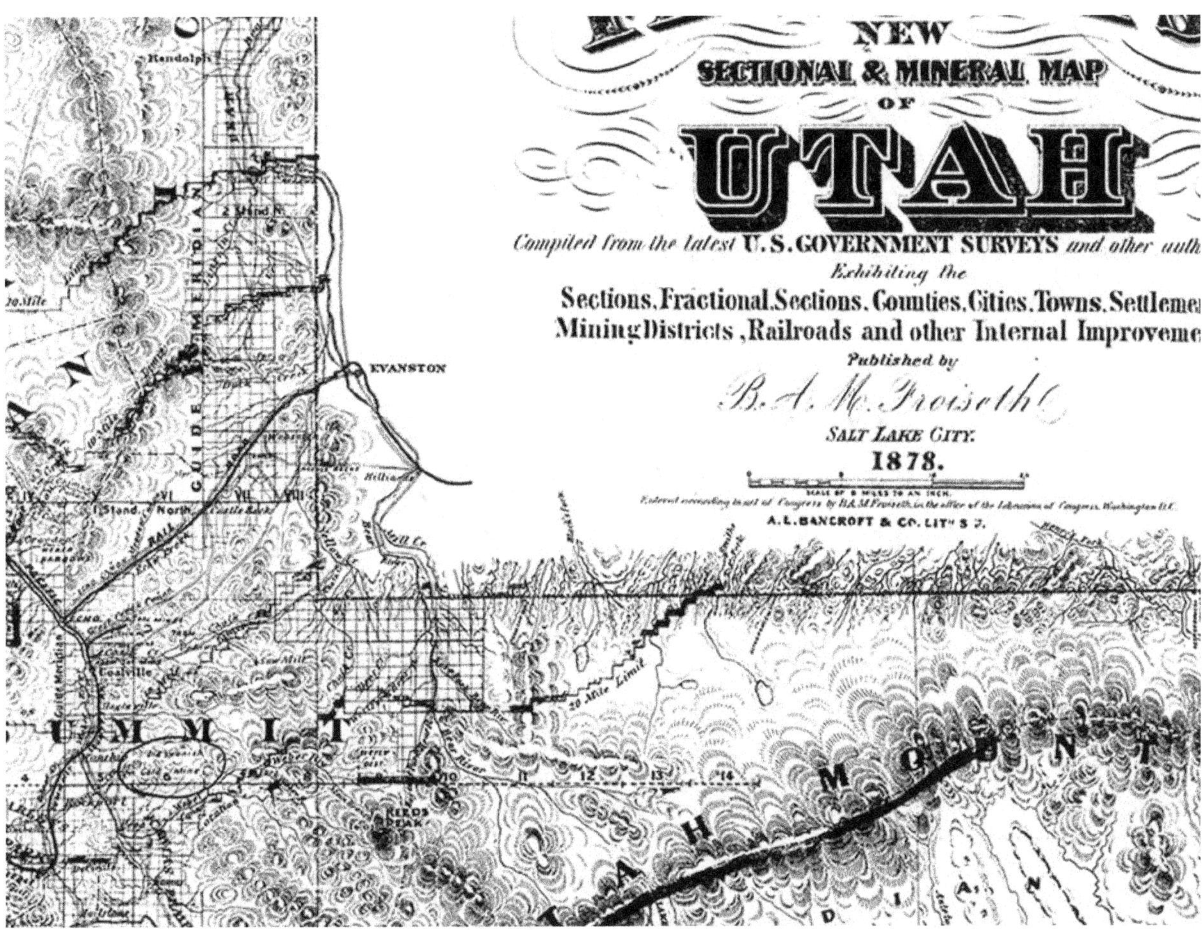

This 1878 Froiseth's mineral map of Utah shows a circled area that says: Old Spanish Gold Mine. (I circled the words on my copy to point them out more clearly.) Upon closer examination, it is difficult to determine where the referenced mine is exactly. The words on the map are located north of Kamas and north of the Weber River.

One of the copies of this map I found years ago at the Brigham Young University Library was actually in color (the mineralized areas of Utah were designated a color for each mineral). There are seven small ovals near the gold mine label, and they were colored in as yellow. That is, I presume, in reference to that area being mineralized or containing gold and not areas of placer gold.

Something to be considered here is the concept that this documented mine was old in 1878. This means they must have been in existence long before the pioneers reached this territory. How old they were when first discovered by the pioneers is unknown.

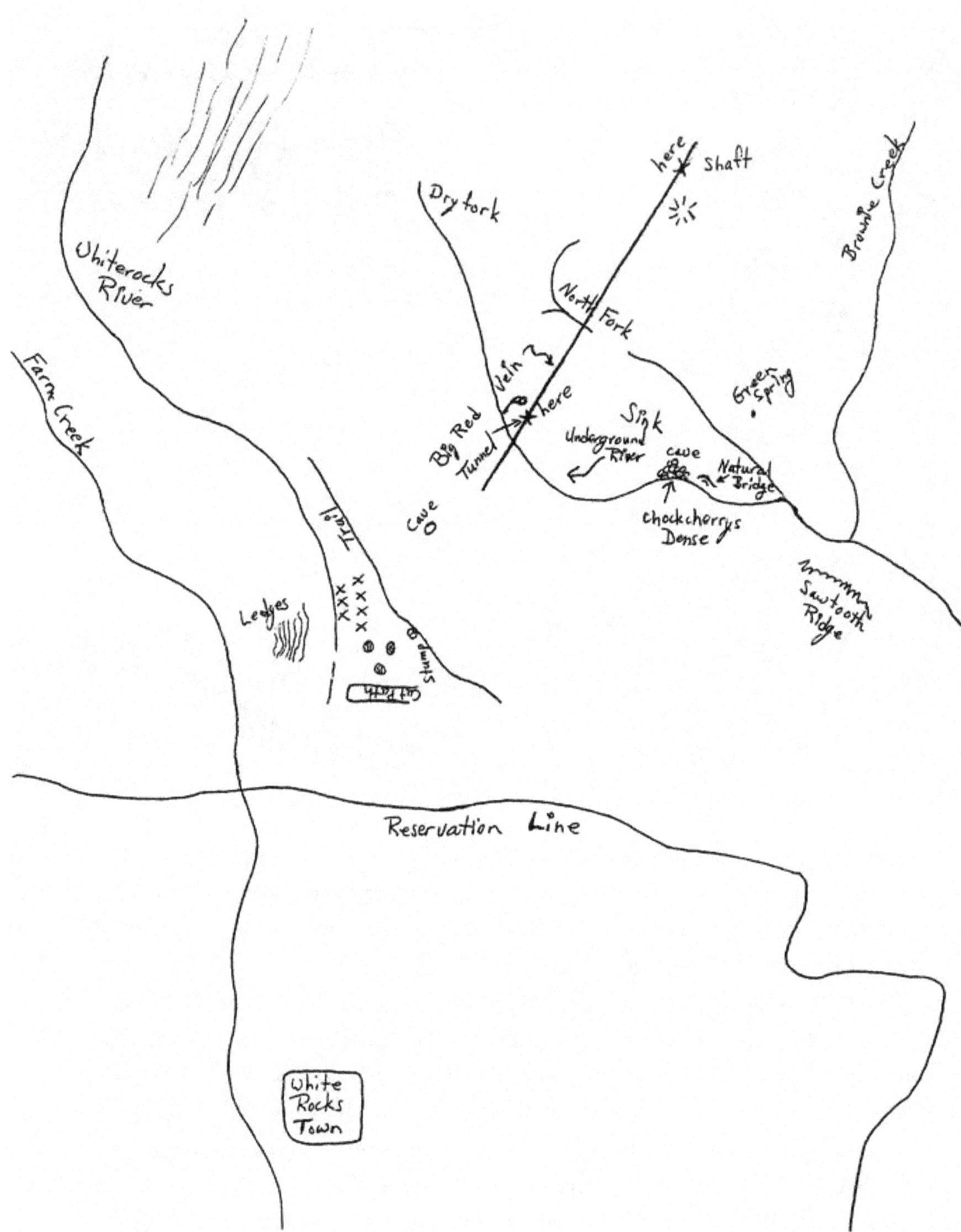

Here is a map of the area from Dry Fork Canyon to Whiterocks Canyon. Verl and another friend gave it to me. The friend told me he had a mine in Whiterocks Canyon which had the same type and color rock as the vein shown in Dry Fork because it was a continuation of the same vein.

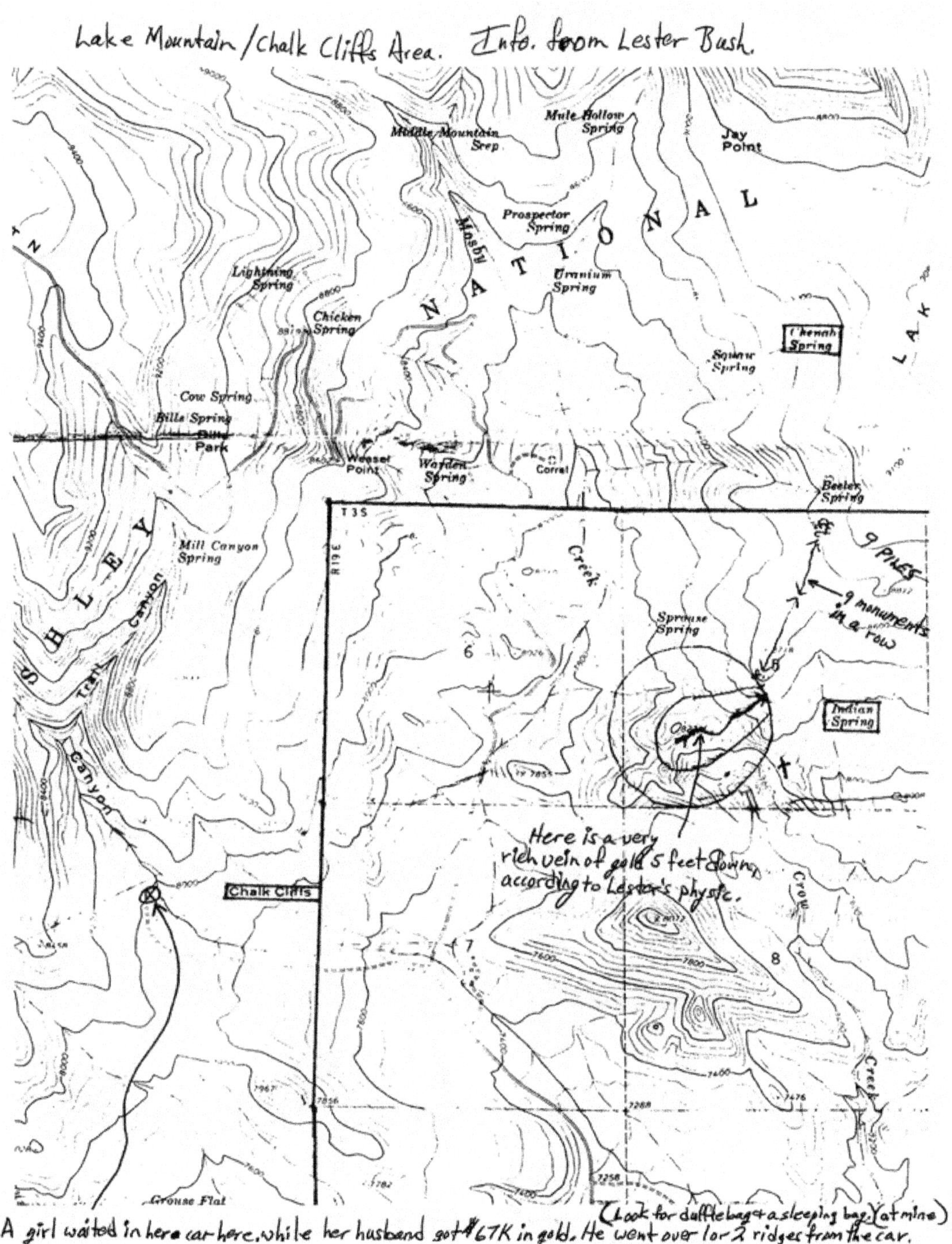

Lester Bush from Oklahoma is the source of the information on this map. George Thompson called him "Okie" in our correspondence. Lester once told me a story about a lady who waited for her husband while he obtained $67,000 in gold from the base of Mosby Mountain. Ken Lance and I couldn't find any evidence that the story was true. However, I believe the vein of gold, which the psychic dowsed and marked for Lester, does exist because it correlates so closely with several other gold stories I've heard about the Crow Creek area. (Hopefully I will remember to include these stories in my next book.)

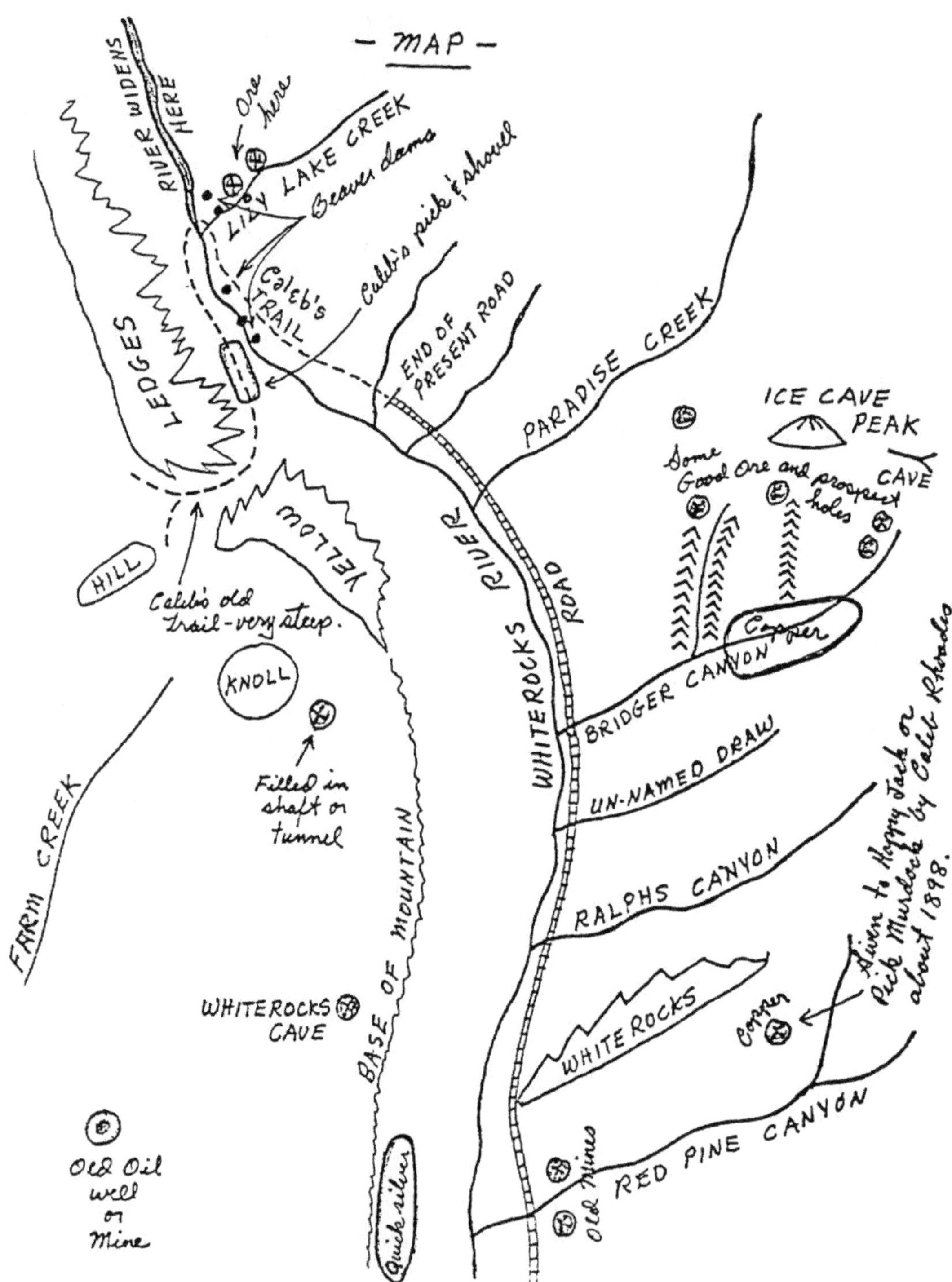

This map is typical of Gale's drawings. One of his strengths was his documentation. He appears to have sketched drawings to document each of his exploring trips. Ed considered this map valid and even added a little bit to it. The word *copper* in Bridger Canyon is in Ed's handwriting.

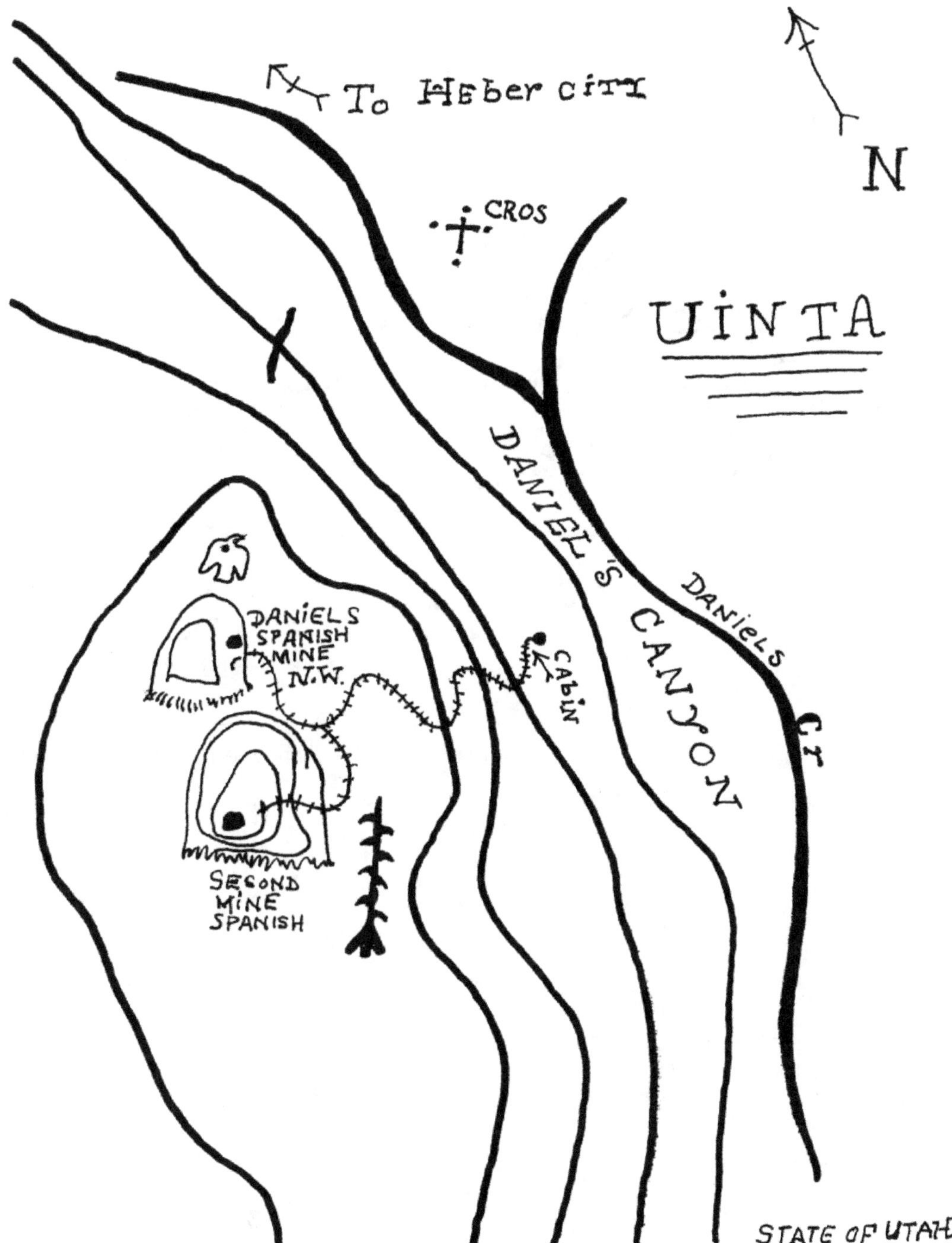

This map is from the bishop. It shows two mines in the Daniel's Canyon area. This is another map that gives directions just as the bishop indicated: Always follow the eagle symbols and not the crosses when you want to find the richest mines.

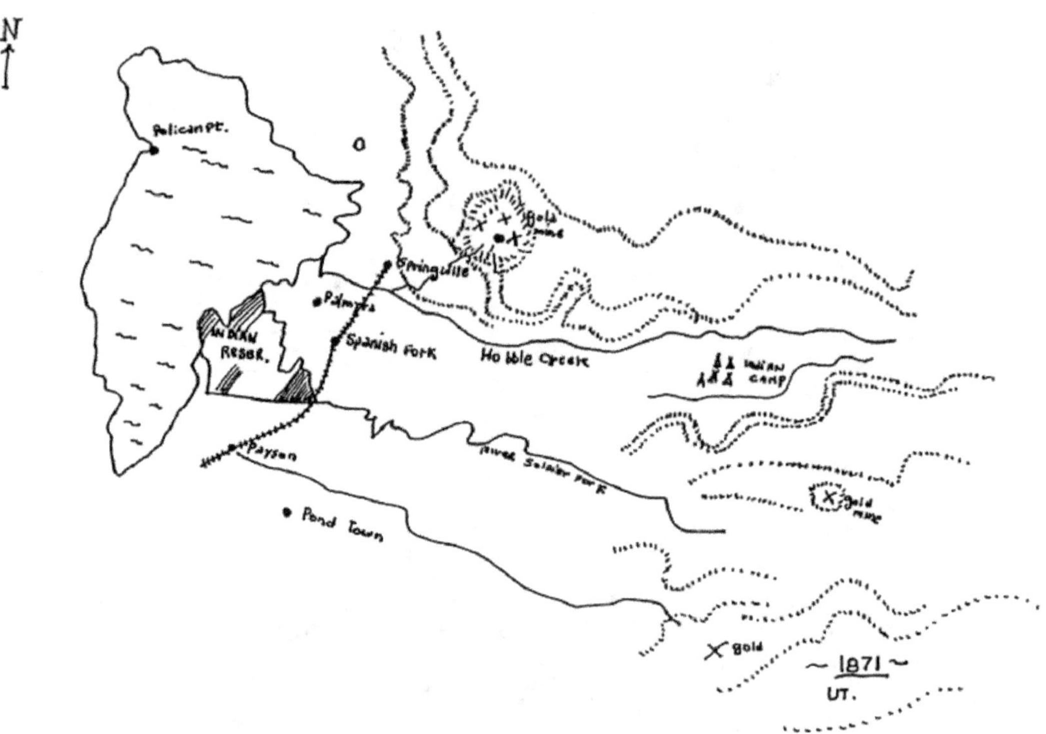

This map shows the Springville/Hobble Creek area.

This is a depiction of an inscribed stone that was said to be about one foot in length and was found close to the location of the present day golf course in Hobble Creek Canyon.

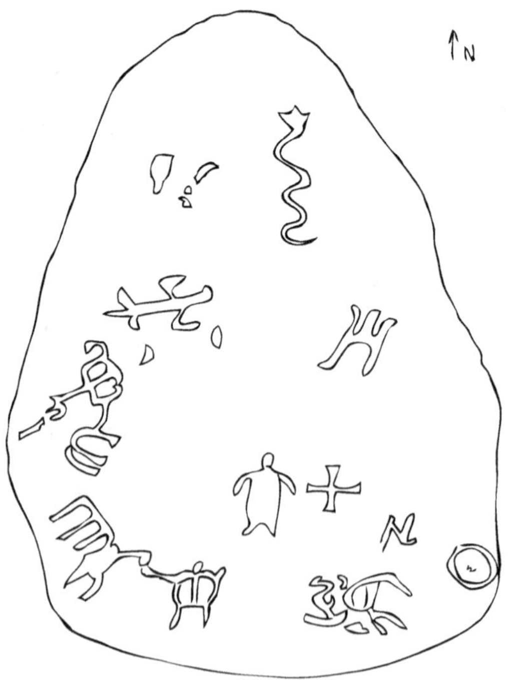

Cris and I purchased this map from the bishop. It may pertain to the Utah Valley/Rock Canyon area near Provo, Utah. It is believed to be a treasure map to 2000 Spanish Reales.

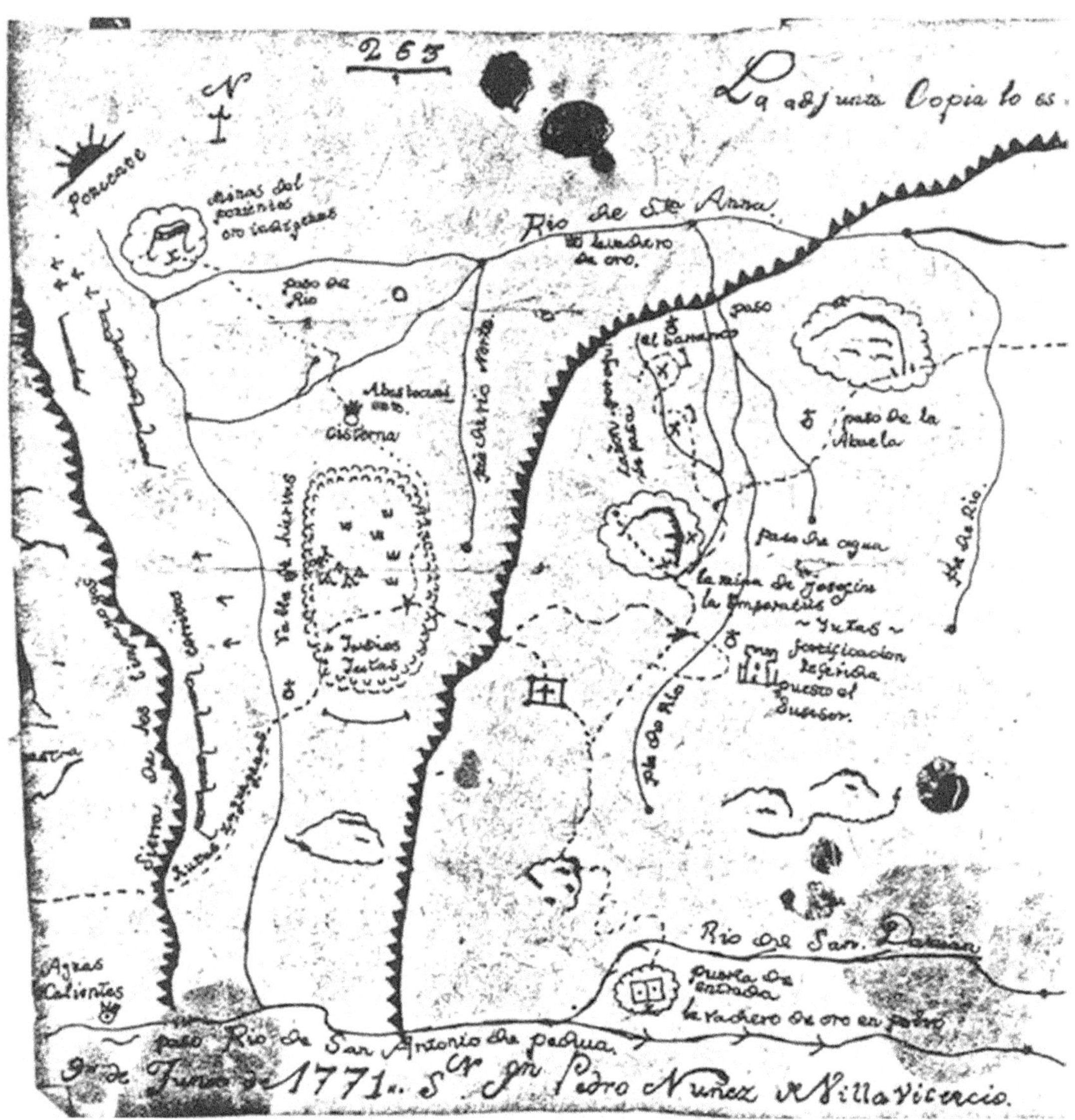

Cris and I purchased this 1771 Nunez map from the bishop. He gave us permission to place it in a book if we ever wrote one. It is of the Hoyt Peak/Kamas Valley area. He told us the rest of the map or that section to the right of this section is in the archives in Seville, Spain. This is a copy of the left or west side. The translation of this map is on the next page, courtesy of Dan Lowe.

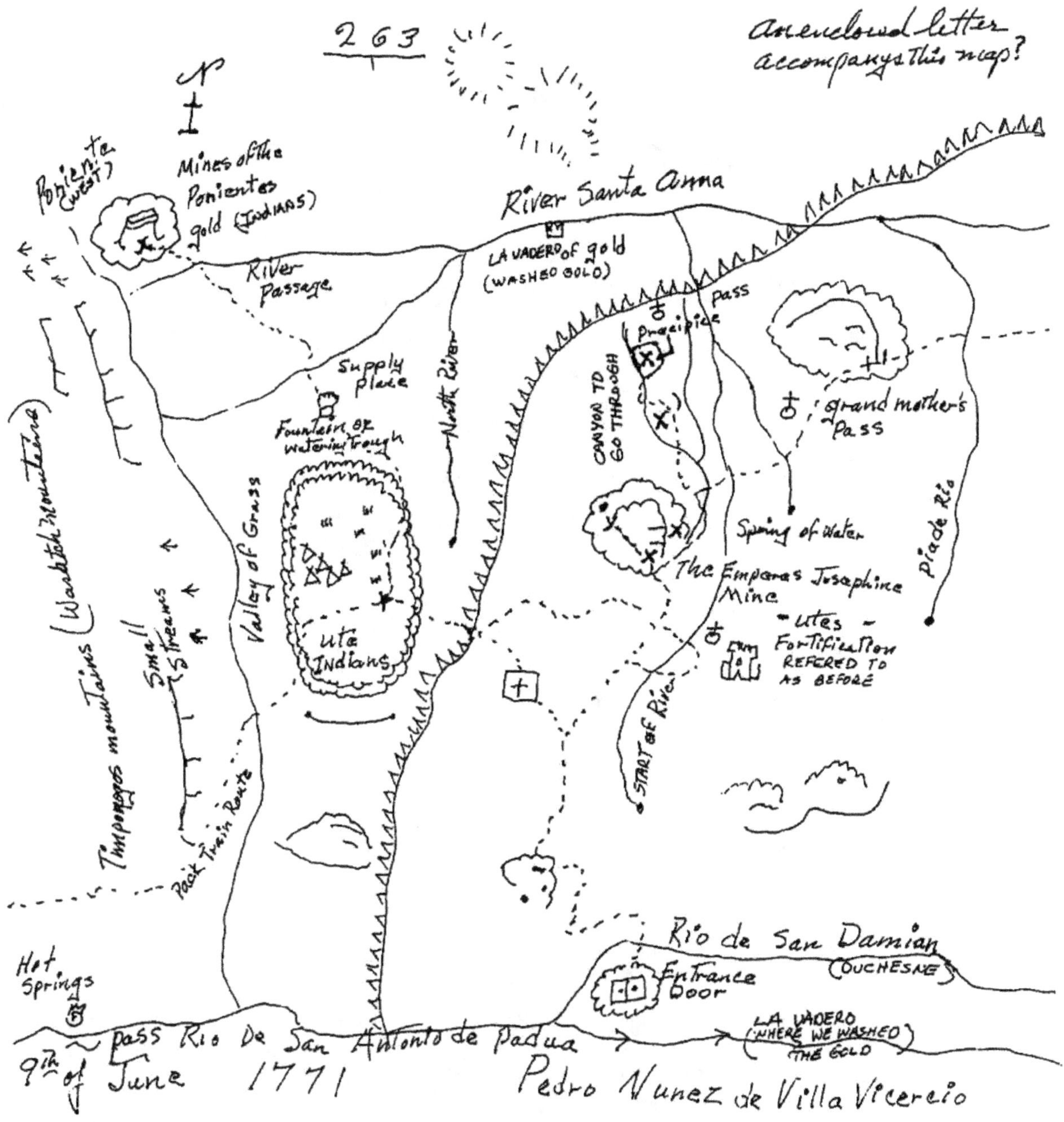

The information above is the English translation of the 1771 Nunez map.

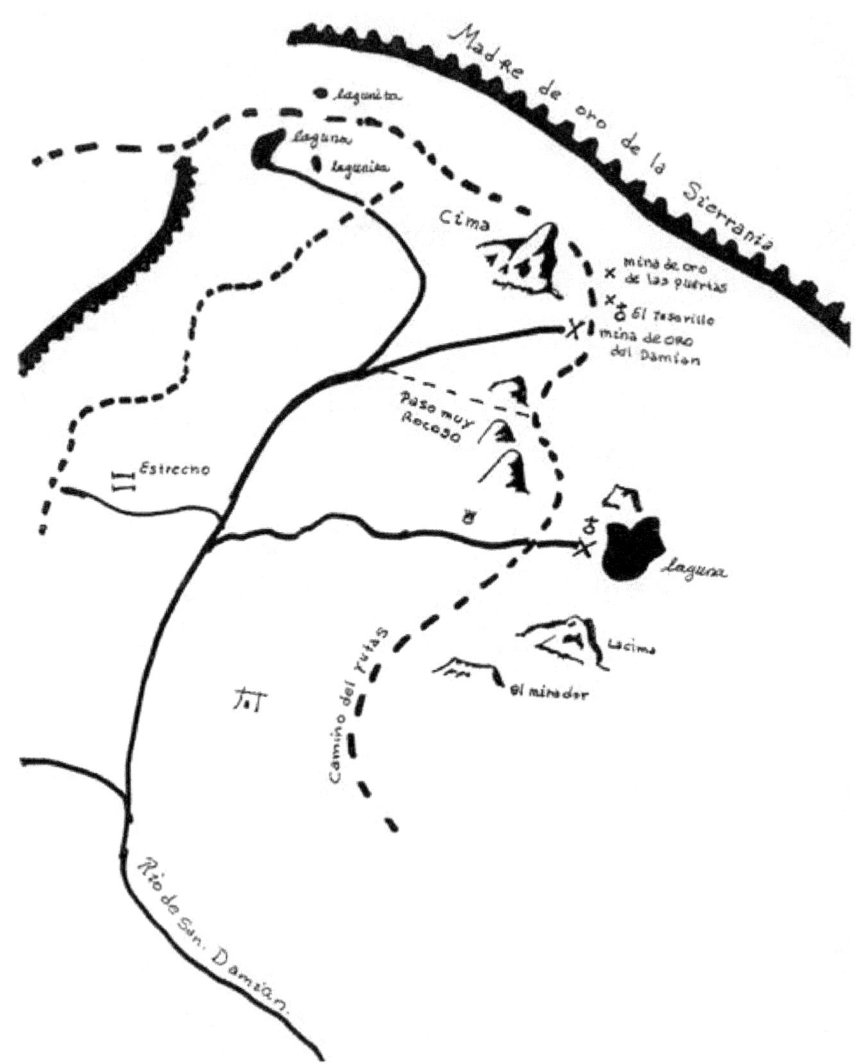

This is a hand-drawn copy of a small section of a 1741 Peralta map. Usually the Peralta name is associated with mines in the Superstition Mountains of Arizona. I have not yet ascertained whether these names and maps correlate in any way. According to Aaron, there are six known copies of this particular Peralta map. Allow me to describe it to you. It is a leather map that rolls up and slides into a silver cylinder with leather caps. There is Spanish writing on the side of the cylinder. On the left edge of the map are mines that were worked by the Spanish west of the present-day Great Salt Lake. On the far right end are mines in what is now Colorado. It shows more than one hundred mines that they worked up to that point. The later Reinaldo 1851–1853 map bears a striking resemblance to this earlier Peralta map. Take a peek at the next map to see the similarities.

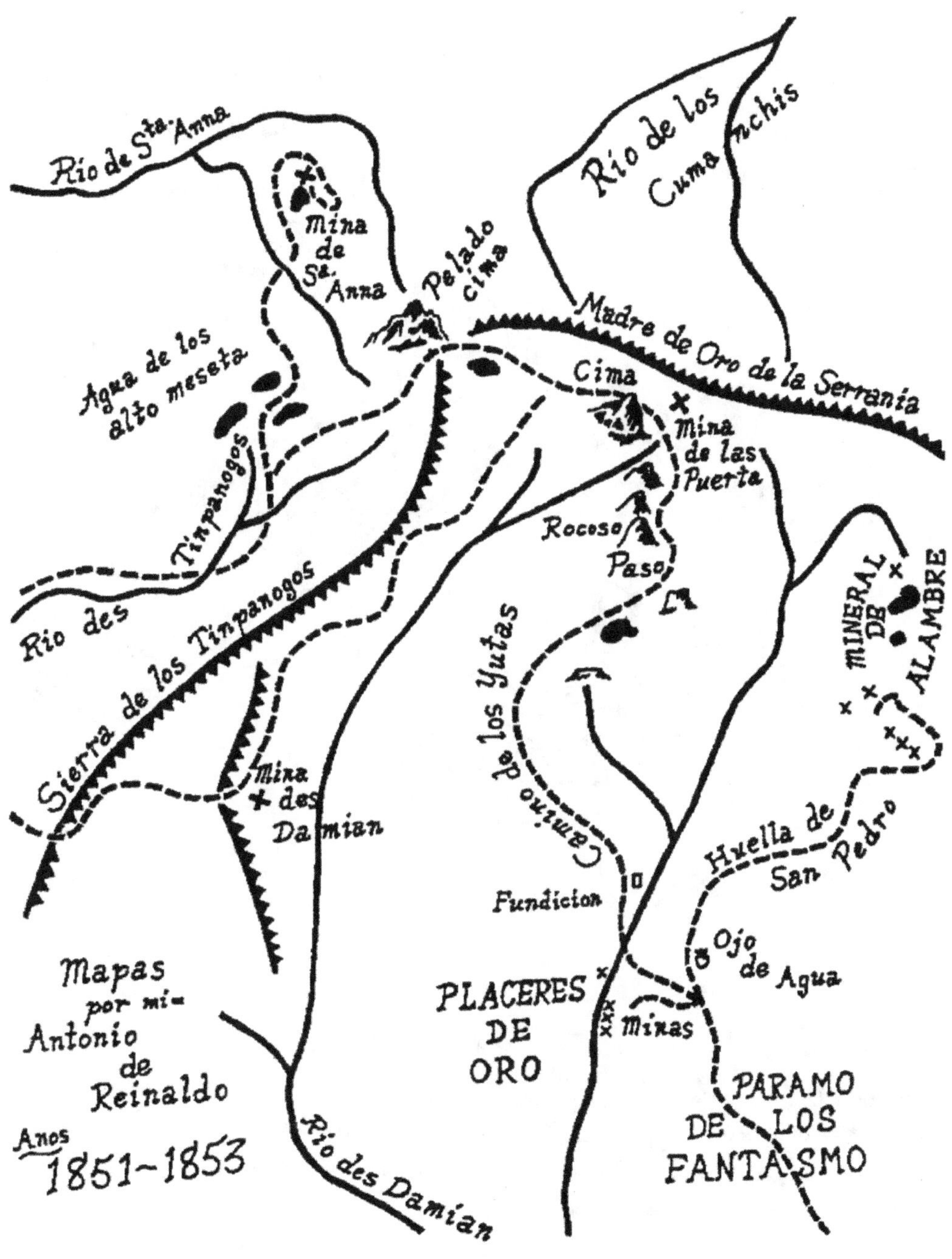

This is the center section of the Reinaldo map. Notice any resemblance to the Peralta map on the facing page?

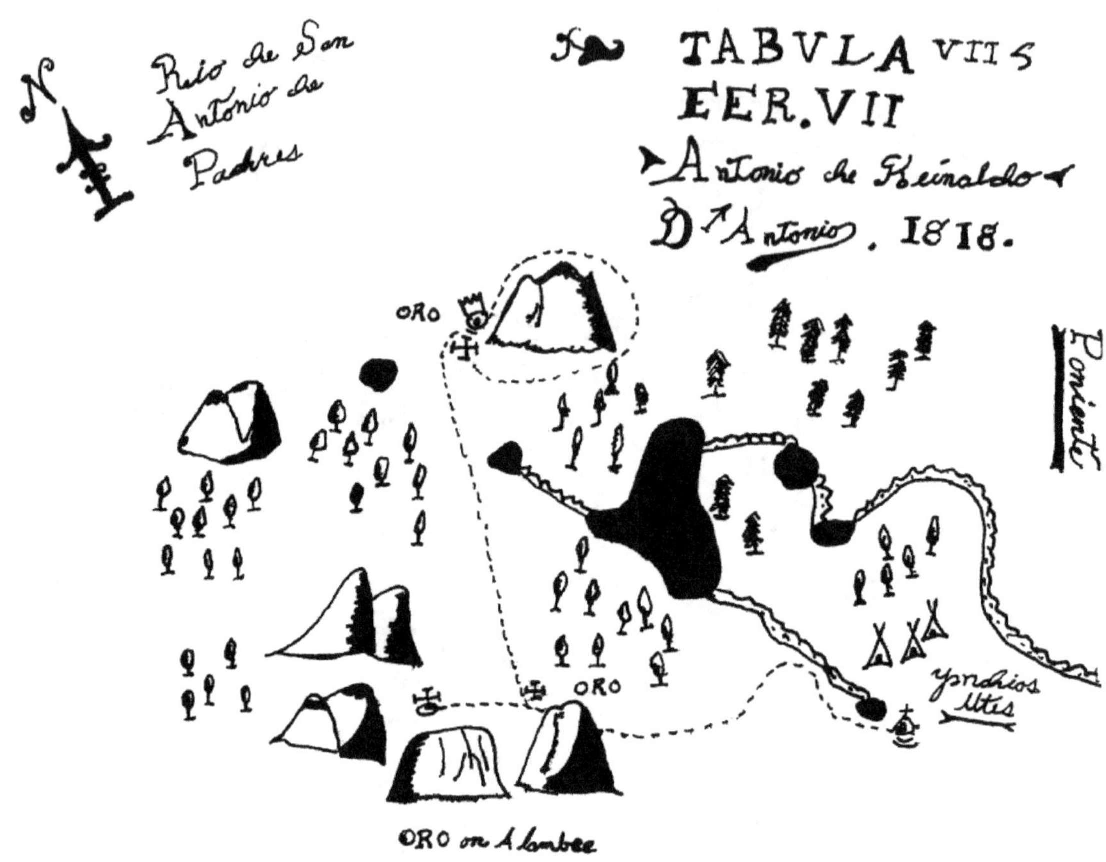

~ Explicacion del oro y plata que se encuentran
en lo Alto de la Meseto del Winty
y se encuentran muy escondidos entre muchos
Lagos, Aqui en este sitio se encuentran mucho
mineral de oro y plata de alambre con cristal
de rocas veteado, Muy Bueno el Lugar ~.

Translation: Explanation about the gold and silver which are found in the high mountain meadows of the Uinta (Winty). They are hidden between many lakes. Here there is a rich abundance of wire gold and silver in quartz crystal veins.

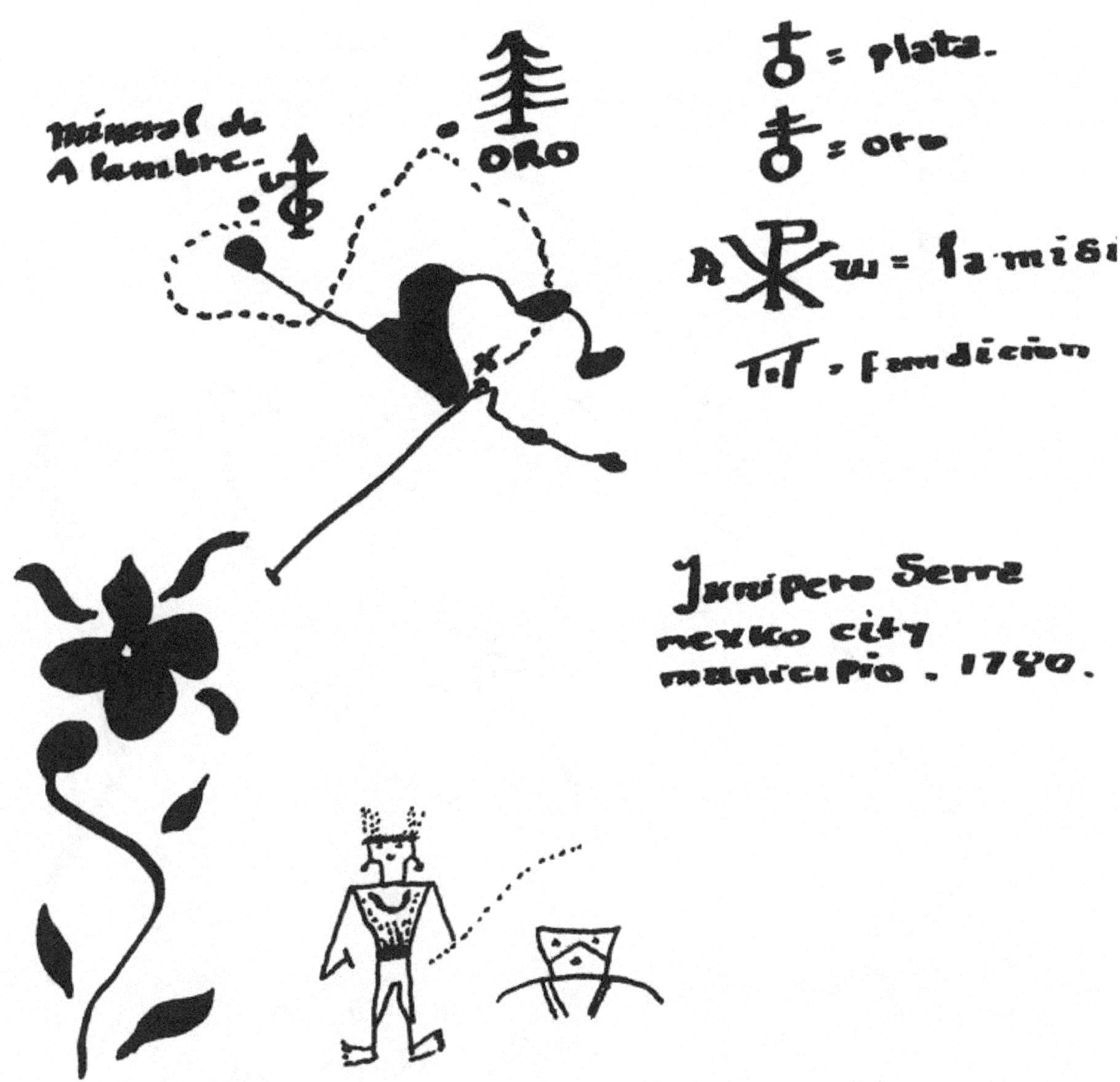

I believe this 1780 map is authentic. It appears to illustrate the Kidney Lake area.

This is a copy of one of the Reinaldo maps found at the Chicken Creek massacre site. Brigham Young sent a U.S. Marshall, who happened to be the bishop's wife's ancestor, to investigate the incident. I obtained my copy of the map from the bishop. The original measures about four inches by five inches. This copy has been enlarged for easier reading.

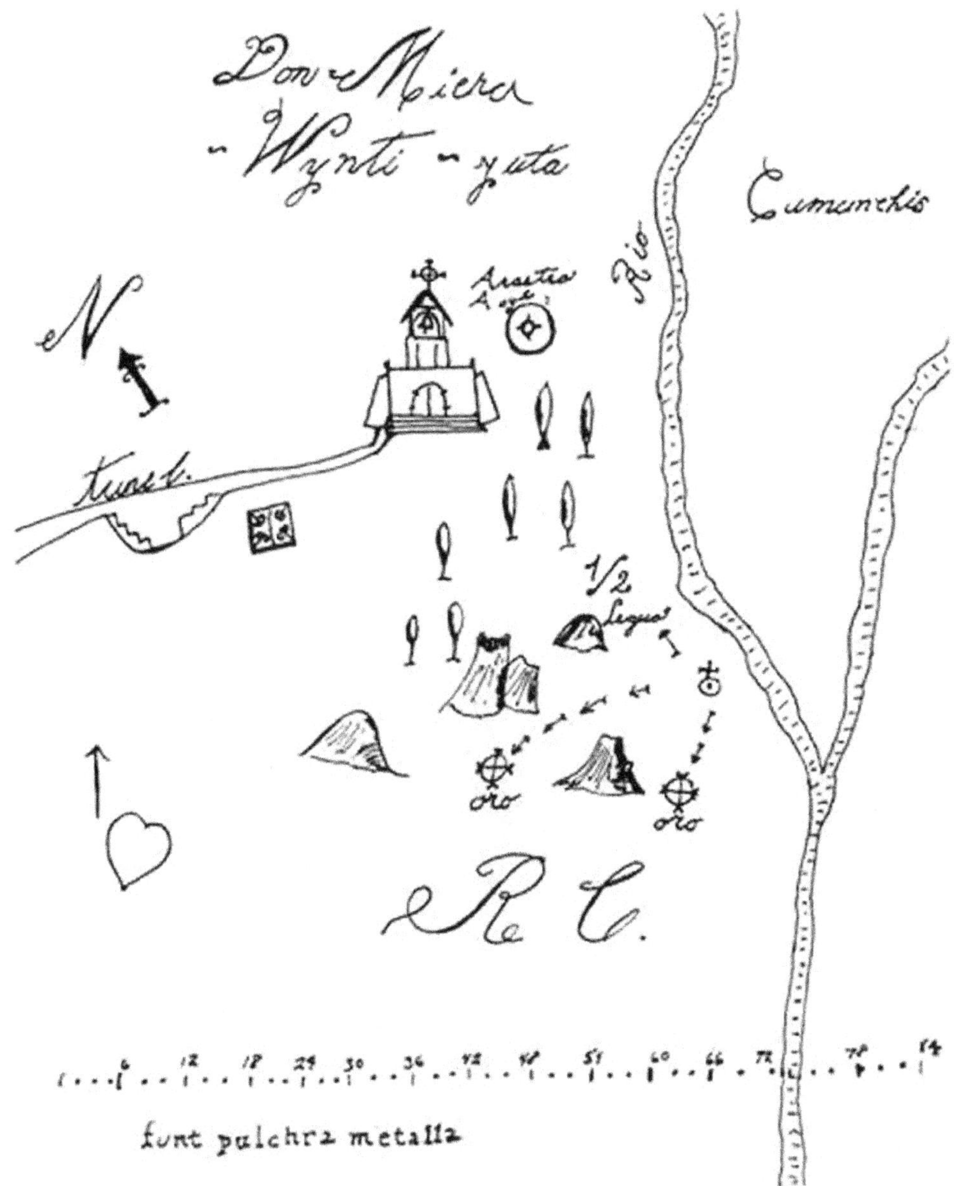

Wynti and Yuta . . . that sounds interesting. The old and authentic Spanish maps seem to describe the Uinta Mountains with these names. Notice the Spanish geographical baseline at the map's bottom. I still have much to learn about their mapping system. This same type of cartography was also on the 1771 Nunez map. The baseline number is at the top of the Nunez map. As we learn about this intriguing history and seek for answers, it's important that we remind ourselves that it isn't so much the destination as it is the enjoyment of the journey that we seek.

Fact or Fiction?

Oro means *gold* in Spanish. This tree is in the Moon Lake area. An arrow is below the word and points toward Moon Lake.

The Spanish were here not just to trade for Native American slaves, as is acknowledged in the history books, but they were discovering, marking, and revisiting their gold and silver mines too. Even before Reinaldo in 1853, Peralta was here and was documenting the New World mines, of which there were over a hundred shown on his map of 1741. Hopefully by the second edition of this book I will have photographed and documented this leather map. If this is accomplished, I will give you the details of it in that printing.

This photo is courtesy of Bill Fink.

Did Caleb Baldwin Rhoades have mining claims in the mysterious Uinta Mountains? Did Thomas and later his son Caleb obtain gold for the LDS Church? Was this action sanctioned by Brigham Young and Chief Walker with an Native American named Ridley as their escort?

There have been times when I have questioned and doubted these things. However, the more I study and the more I uncover, the more convinced I become that it really happened just as the legends say it did.

I yearn for the day when we will have all of the puzzle pieces and will put the last one in place. Then and only then will we know the whole story and every exciting tidbit. What an incredible day that will be!

Occasionally as I explore I find evidence to back up the legends. This boulder with Caleb's initials chiseled into it lends credence to the legends. It's located near Hanna, Utah, on Native American land. The initials FWC stand for F.W.C. Hathenbruck, who was Caleb's mining partner. This photo is courtesy of Randy Lewis.

Is there Evidence of Minerals in the Uinta Mountains?

This picture of an igneous dike on the eastern slope of King's Peak is courtesy of Woody Olsen.

Mineralization can be found from the copper-colored rock of the south flank fault at Swift Creek and the south fork of Rock Creek to the igneous dike said to run from the Canyon of Ladore to Gilbert Peak. At several locations in the Uinta Mountains, iron and native copper have been found and documented.

The Spanish called the Uinta Mountains the Madre del Oro or the mother of gold. While it is difficult to pan the streams and find significant amounts of gold consistently, the possibility still exists. Blind Steam, for example, is on the state's list of streams accessible for dredging.

While exploring near Moon Lake during the summer of 2004, I stumbled upon this colorful outcropping of peacock copper ore. It was probably deposited when hydrothermal solutions pushed through the cracks in the host rock and deposited a thin layer of minerals. It's on the embankment on the south side of the stream where Fish Creek joins Moon Lake. The same basic ore with slight variations may be found running mainly east and west at various locations throughout the Uinta Mountains.

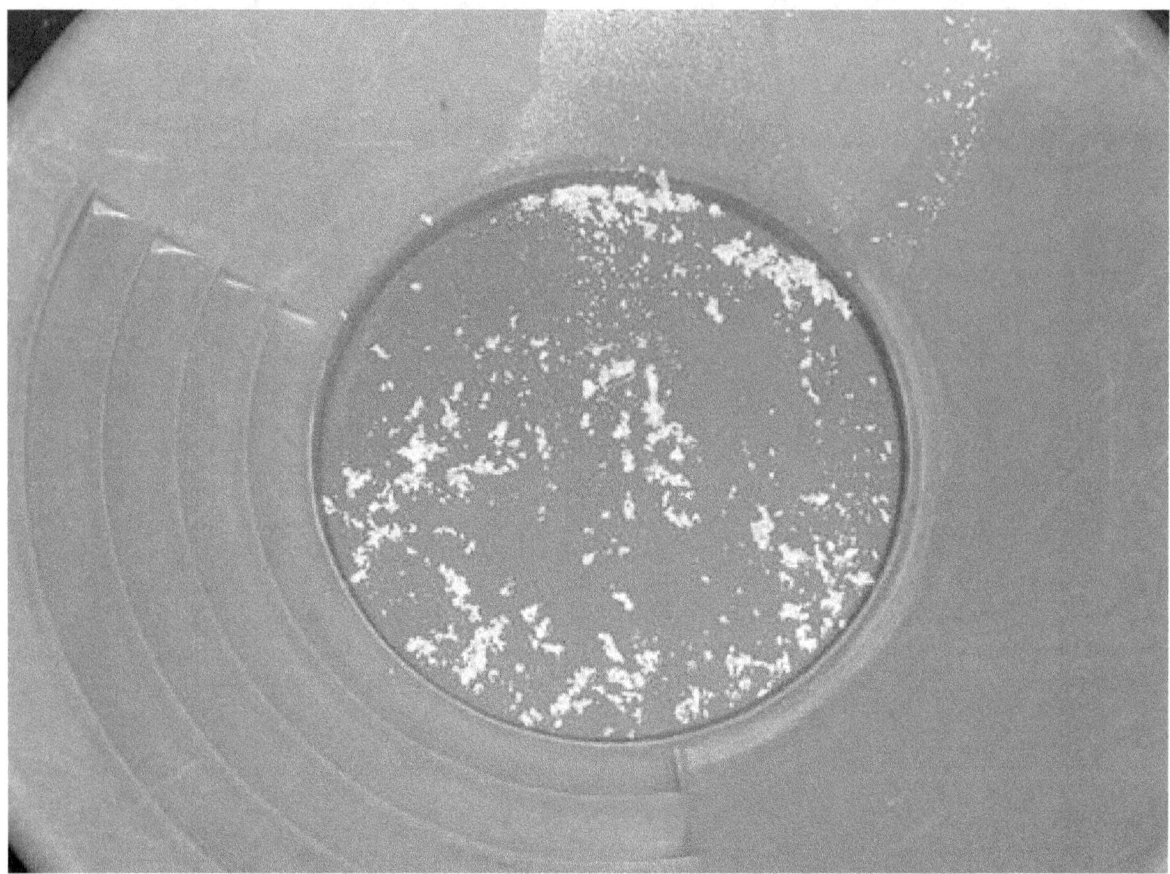

The gold pictured here is about one-fourth of the gold he ended up with after crushing a rock the size of a small football.

Is there any gold in the Uinta Mountains? That's an excellent question. Over the years, as I have panned and examined rocks from various canyons in the Uinta Mountains, I have never personally located any significant quantities.

Others have not had the same results. For example, the gold in the pan pictured above was given to me by a friend named Steve Warby. He stumbled upon a rock in Dry Fork Canyon which looked unusual to him and upon closer examination discovered it contained sheet gold. He crushed the rock and gathered up the gold released from its cracks.

There are locations in Current Creek and Raspberry Draw where extensive placer operations took place near the turn of the century. It is documented that the copper on Dyer Mountain had a significant amount of gold in it. (See the "Precious Metals at Dyer Mine" chapter.) Also, the intrusive dike near Deadman Lake reportedly contained gold and nearly caused a stampede around 1900. (See the "Hundreds of Claims at Deadman Lake" chapter.) There are many more stories and legends yet to be verified that lend credence to the theory that significant quantities of gold do exist in the mysterious Uinta Mountains.

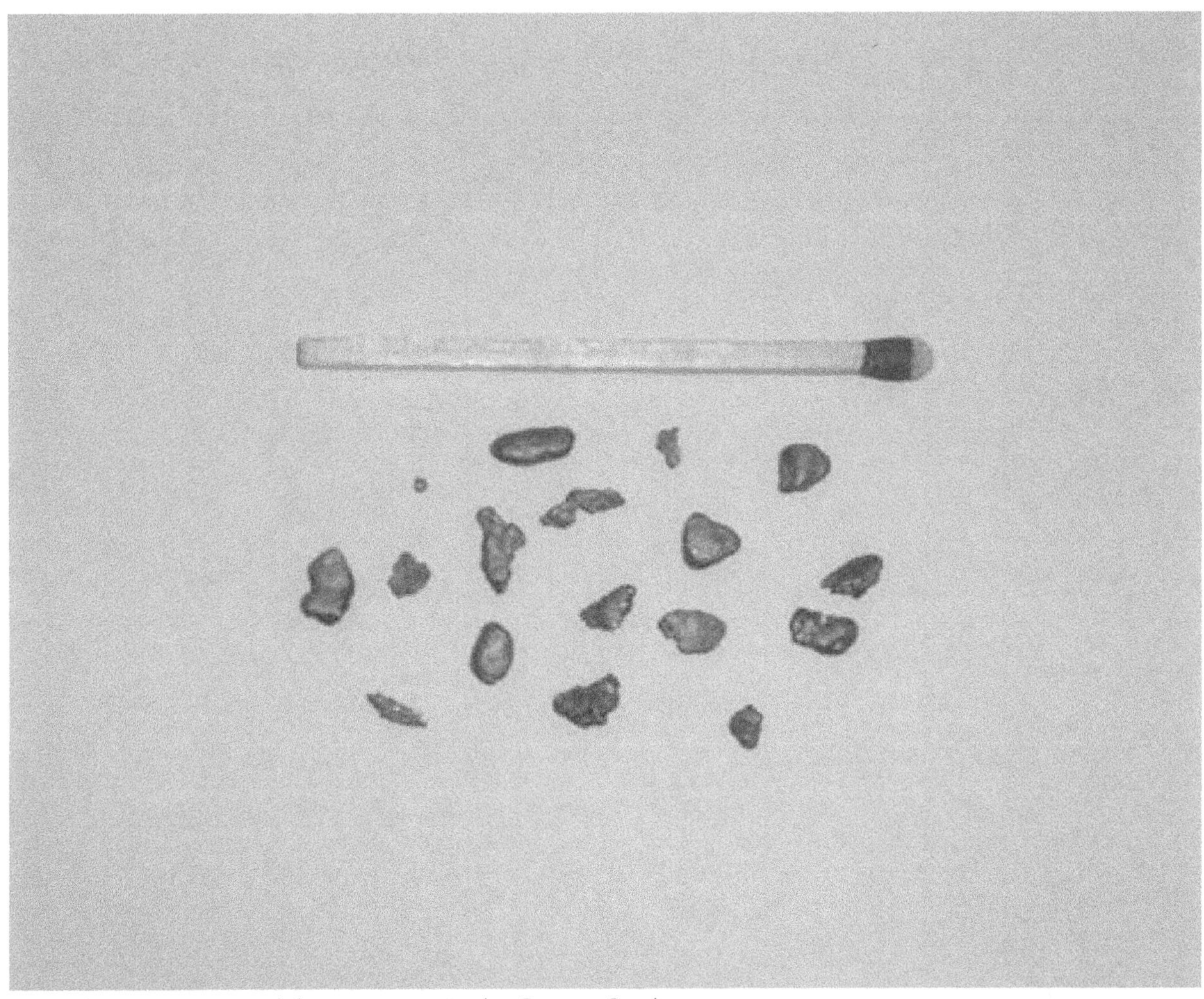
Gary Christensen panned these nuggets in the Current Creek area.

This picture helps to verify the existence of placer gold in the Uinta Mountains.This is a small cluster of nuggets that I purchased from Gary Christenson. One day Gary called me and said he needed some fast cash for expenses. He offered me these nuggets. I jumped at the chance.

Even though they are small, they add up to about one quarter of an ounce. Gold is very heavy. The largest nugget is about the size of a match head. Thanks, Gary, for lending a little more credibility to the legends.

Epilogue

Hopefully I have given you something to consider. If nothing else, maybe it has been entertaining. I do not aspire to convince you of the feasibility of this work or of its validity. My passion has been to bring this information to light and not to see it lost. I have only a small desire to convince you of the truth of what I have presented. It has been said that a man convinced against his will is of the same opinion still.

Writing this work hasn't been an easy task. I have walked a thin line. There are those who would have me say nothing and keep it all for themselves or they are worried I would say too much about this sacred gold and offend God or the Native Americans. Others have yearned for every morsel and tidbit. It isn't my position to judge any of them. I have done my best, with my conscience as my guide, to present this information as accurately as I possibly can. Hopefully I have walked the middle ground and have pleased all sides without offending anyone.

Yes, I have been blessed and privileged to catch a glimpse of Utah's rich history. It's a history that speaks of men's dreams and aspirations. It tells a story of battles, of enslavement, and of man's quest for power through riches. In the process of research and interviews, through hiking the hills and pausing every now and then long enough to see what is really there, the clues have occasionally jumped out at me and the puzzle pieces have fit into place. In fulfilling this process it has been my privilege to feel at one with my quest and to come to the understanding that these things really did happen as the legends say they did.

Our Maker has a time and purpose for these riches. I do not know what that purpose is. Hopefully I will never be arrogant enough to think He could not succeed in His plan without me or what I have gleaned. His plan will continue with or without my assistance. The riches of gold may never be mine. However, I feel rich beyond measure for the friendships I have made and the opportunity I have to touch the past as I follow the legends . . .

Bibliography

Johnson, Stan and Polly. *Translating the Anthon Transcript*. Parowan, UT: Ivory Books, 1999.

Kenworthy, Charles A. *Spanish Monuments & Trailmarkers to Treasure in the United States*. Encina, CA: Quest Publishing, 1993.

Mathewson, R. Duncan III. *Treasure of the Atocha*. Key Largo, FL: National Center for Shipwreck Research Ltd., 2004.

Pickett, Mike "Hawkeye." *Treasure Hunter's Field Notebook*. Ceres, CA: THU Publishing Company, 2001.

Rhoades, Gale R. & Boren, Kerry Ross. *Footprints in the Wilderness, A History of the Lost Rhoades Mines*. Salt Lake City: Dream Garden Press, 1980.

Rhoades, Gale R. *Lost Gold of the Uintah, the Rest of the Story*. Salt Lake City: Benzoil Inc., 1995.

Rhoades, Gale R. *Waybill to Lost Spanish Mines and Treasures*. Salt Lake City: Dream Garden Press, 1982.

Uinta Mountain Geology. Salt Lake City: Utah Geological Association, 2005.

Index

About the Author

Dale Rex Bascom was born in Riverton, Wyoming. When he was a year old, his family moved to Idaho, where they homesteaded 148 acres on a G.I. bill in the Snake River Valley. There he enjoyed finding arrowheads on scout hikes and exploring a local lava-flow cave where Native Americans had once lived. He also found his first dinosaur bone on the farm and has been hooked on rock hounding ever since.

In 1965 the family moved once again, this time to Spring Lake, Utah. With the close proximity of the mountains, he often hiked the hills and explored the nearby mines.

While growing up, his hobbies included fishing, deer hunting, and hiking. Dale completed a two-year mission in Germany for The Church of Jesus Christ of Latter-day Saints in 1974. Upon his return, he attended what is now UVSC, where he graduated with an associate's degree in electronic technology. He met his future wife at school, and they were married in the LDS Manti Temple in 1975. Dale and Jeri have six children and will shortly have ten grandchildren. He has been employed at Brigham Young University since 1984.

www.ingramcontent.com/pod-product-compliance
Lightning Source LLC
LaVergne TN
LVHW081251100826
845148LV00009B/1190

* 9 7 8 1 5 9 9 5 5 0 4 3 5 *